Psychophysics of Reading in Normal and Low Vision

Psychophysics of Reading in Normal and Low Vision

Gordon E. Legge
University of Minnesota

CRC Press
Taylor & Francis Group
Boca Raton London New York

CRC Press is an imprint of the
Taylor & Francis Group, an **informa** business

CRC Press
Taylor & Francis Group
6000 Broken Sound Parkway NW, Suite 300
Boca Raton, FL 33487-2742

First issued in paperback 2021

© 2007 by Taylor & Francis Group, LLC
CRC Press is an imprint of Taylor & Francis Group, an Informa business

No claim to original U.S. Government works

ISBN-13: 978-0-8058-4328-6 (hbk)
ISBN-13: 978-0-367-39014-3 (pbk)

Visit the Taylor & Francis Web site at
http://www.taylorandfrancis.com

and the CRC Press Web site at
http://www.crcpress.com

To Wend and Alex with love,

—Gordon E. Legge
July, 2005

Contents

The accompanying CD contains full reprints of the twenty original articles in the *Psychophysics of Reading* series by Gordon E. Legge and his colleagues. See Box 1.1 for a list of these twenty articles. The CD also contains a cumulative reference list with all the citations from the book and the twenty articles.

CONTENTS

PREFACE and ACKNOWLEDGMENTS

On a cool, gray day in October 1976, I arrived at the Craik Vision Laboratories at Cambridge University. I had passed my PhD defense at Harvard only a few days earlier. I was about to begin the most important year in my research career— a postdoctoral position in the laboratory of Fergus Campbell.

I first met Fergus on December 12, 1973 when he gave a lecture at Harvard University entitled, "Fourier Analysis and Visual Perception." At the time, I was a graduate student in psychology, but because I had an undergraduate degree in physics, I was familiar with Fourier theory. Fergus's lecture stimulated my interest in the spatial-frequency analysis of vision, and soon I was at work on a thesis in this area. Subsequently, David Green, then at Harvard, encouraged me to do post-doctoral research at Cambridge, and wrote to Fergus on my behalf. Even though my first published article (Legge, 1976) included a critique of the Fourier model, Fergus agreed to sponsor me.

Within a day or two of my arrival at Cambridge, I met several of the giants in the field, including William Rushton, Mat Alpern, Horace Barlow, and John Robson. During my postdoc year, I interacted with and learned from many students and sabbatical visitors at Cambridge who were, or would soon become, major stars in the field, including John Foley, Howard Howland, V. S. Ramachandran, Andrew Parker, Dave Heeley, Art Ginsburg, and Jan Atkinson. Denis Pelli was my office mate. Denis and I had met previously, when he was an undergraduate at Harvard and I was a graduate student, but our time together at Cambridge cemented a life-time friendship.

Toward the end of my year at Cambridge, John Ross, a visitor from the University of Western Australia, demonstrated a new type of visual display he'd invented called the betagraph. It consisted of several widely spaced vertical columns of light-emitting diodes (LEDs). By appropriately sequencing illumination of the LEDs, the display simulated the presentation of drifting text, as if seen through the gaps in a picket fence. For an observer who employed proper pursuit tracking eye movements, highly legible text appeared to drift across the face of the display. The phenomenon of apparent motion ("beta" motion), made the drifting image appear continuous, although depicted by a sparse array of sampling elements. The large size of the letters on the display, their depiction by bright elements on a dark background, and the salience of the motion, all made the display visually compelling for me, a person with impaired vision. The betagraph demonstrated that legible text could be displayed with very few picture elements at a relatively high magnification. Both John Ross, the inventor of the betagraph, and Fergus Campbell, who was always interested in clinical applications of vision research, encouraged me to consider the betagraph design as a low-vision reading display.

Soon after seeing the betagraph, I moved to the University of Minnesota as a new assistant professor. I set up a lab to study spatial-frequency masking, contrast discrimination and binocular interactions. I obtained a grant from the National Institutes of Health (NIH) and soon had a fancy computer system (DEC 11/23) that could generate sine-wave gratings and related one-dimensional luminance profiles. I had a good start on my research career.

But I continued to think about the concept of sparsely-sampled matrix displays for low-vision reading. I thought about how such displays could be simulated for laboratory study. I made some inquiries with local foundations in the Minneapolis-St. Paul area, hoping to obtain modest funding for a secondary project. When these inquiries failed, I decided to submit a small two-year grant proposal to NIH. I recognized that an NIH submission would require demonstration of knowledge of the relevant literature on low-vision reading, such as the effects of character size and field size. To my surprise, I discovered there was almost no psychophysical research on reading and low vision. Moreover, the psychophysical work on reading and normal vision was not well-integrated with modern concepts from vision research. Miles A. Tinker and his colleagues, coincidentally also at the University of Minnesota, had conducted a series of studies on the legibility of text from the 1920s to the 1950s (reviewed in Tinker, 1963). This work was conducted prior to the major advances in spatial vision in the 1960s and 1970s. Tinker focused primarily on factors affecting the "legibility of print," with almost no attention to the underlying visual mechanisms. Tinker's work, while a treasure trove of empirical findings, left many key issues hanging. For example, his studies of print size typically did not control for viewing distance, so conversion to angular character size, an influential variable for reading speed, was not possible. Moreover, Tinker did not extend his research to impaired vision. It was soon evident to me that my small NIH grant would have to include exploration of the effects of some very basic stimulus variables.

Never underrate the importance of good luck. My small grant entitled, "Matrix Displays in Low Vision Reading Aids" arrived at NIH at the time when the National Eye Institute (NEI; part of NIH) wanted to initiate research on low vision. My application may have been the first fundable grant in this area. They made the decision not only to fund the grant, but to double its budget to include a post-doc. The reasoning was that the topic was of sufficient importance to merit a greater commitment than I had requested. But whom should I recruit for this position? I knew Denis Pelli was nearing the end of his PhD at Cambridge, and I also recognized his creative genius. It was yet another stroke of great good fortune that he agreed to come to Minnesota to work with me on reading and low vision.

Gary Rubin and Dan Kersten were my first two grad students at Minnesota. Gary came to vision research from cognitive psychology. He had an immediate appreciation for the significance of reading as a topic, and an interest in the low-vision applications. Dan was a subject in the initial experiments, and contributed insightfully to our formative plans. Mary Schleske rounded out our initial lab group. She drew on her experience in nursing to help us develop methods for recruiting and testing low-vision subjects, and for making them comfortable and productive in the research.

In subsequent years, many wonderful students and colleagues have worked with me on vision and reading, and are named as authors in the collection of articles on the compact disk accompanying this book. It has been, and continues to be, a joy to work with them. Of particular note, Andrew Luebker made innumerable technical contributions, especially to the development of computer-based methods for measuring reading performance.

The NEI has continuously funded this research since the first award in 1979 for work on matrix displays. I gratefully acknowledge support from NIH grant EY02934. I am deeply indebted to three NEI program officers for their help, encouragement and friendship over this period of time—Connie Atwell, Israel Goldberg, and Michael Oberdorfer.

I want to thank many people who helped me with the preparation of this book. Steve Mansfield is the first author of Chapter 5 and the appendix on print-size conversions. He also reviewed Chapter 3, and prepared the contrast demonstration in Fig. 3.1. Susana Chung reviewed all of the chapters, provided me with Fig. 3.6, and raised many key points in discussion. Paul Beckmann prepared the index. Josh Gefroh and MiYoung Kwon played major roles in library research, and in creating, refining, and assembling figures, reference lists, and draft manuscript material; I am particularly grateful for their thorough and conscientious attention to the details. Sing-Hang Cheung, now nearing completion of his Ph.D. as my student, prepared many of the new figures and helped with several important calculations. Deyue Yu, also one of my current graduate students, conducted the experiment on pixel sampling described in Section 4.3, and helped with several of the new figures and tables. Providence Rao and Lori Handelman at Erlbaum made production of the book possible.

I am also grateful to the following individuals, listed alphabetically, for their generous help: Ameara Aly-Youssef helped with library research and printing of the final manuscript copy. Aries Arditi reviewed Chapter 4. Charles Bigelow reviewed Chapter 4 and provided many helpful comments on typography, fonts, and text rendering. Dwight Burkhardt reviewed Box 4.2 on pixelized vision. Susan Chrysler directed me to relevant sources on the design and visibility of highway signs. Michael Crossland reviewed Chapter 1. Gislin Dagnelie provided Fig. 4.6 demonstrating pixelized vision, and reviewed Box 4.2 on this topic. Don Fletcher provided Fig. 3.15 on the patient with a ring-shaped scotoma. Chris Kallie helped prepare the summary figure of the Braille code in Box 2.2. Don Kline discussed the influences of age and blur on the legibility of signs. Marie Knowlton pointed me to key early references on the geometry of the Braille cell. Jim Larimer provided me with motivation to do the book, and suggested topics that would be of interest to display engineers. Don Meeker provided me with Fig. 4.1, illustrating new and old designs of highway signs, and commented on the rationale for these designs. Bob Morris reviewed Chapter 4, and commented on fonts, spacing and text rendering on computer displays. Beth O'Brien reviewed Box 2.1 on dyslexia and Box 3.3 on dual-route theory. Cynthia Owsley commented on issues related to aging and reading. Eli Peli reviewed the section on fiberoptic magnifiers in Chapter 4. Denis Pelli reviewed Chapter 3. Ralph Radach helped me find sources on the early history of reading research. Gary Rubin reviewed chapters 1 and 2. Sarah Sass helped with the conversion of the original journal articles into manuscript format. Frank Schieber pointed me to Forbes's research on fonts for highway signs. Bosco Tjan reviewed Chapter 3 and discussed numerous detailed issues about topics in the book. Brenna Vaughn assisted with the proofs. Tony Wilson reviewed Box 3.2 on the wordform area.

I owe special thanks to Wendy Willson Legge whose love, encouragement, practical advice and constant help have been critical to the success of my research career and the completion of this book.

—Gordon E. Legge
July 25, 2005

1

Vision and Reading

Gordon E. Legge

This book is about the role of vision in reading. It describes the influence of physical properties of text on reading performance and the implications for information processing in the visual pathways. It deals with the reading performance of people with normal, healthy vision and also people with impaired vision.

The accompanying compact disc contains reprints of our series of twenty published articles on the psychophysics of reading in normal and low vision, published in the period from 1985 to 2002. These articles are reproduced in their entirety, and are identical to the original articles, apart from stylistic reformatting. Box 1.1 lists the 20 original citations.

The goal of the book is to organize and synthesize the major findings of the work in the series of 20 articles and to place them in the context of other contemporary research on vision and reading.

The topic of vision and reading is of interest to many groups of researchers including those in vision science, psychology of reading, clinical vision research, rehabilitation, display engineering, and human factors. The book is intended to speak to people in all of these groups. Readers should feel free to go directly to the sections of particular interest to them. Although the book is written primarily for people with expertise in one or more of these fields, dedicated students should also find it informative. Keep in mind that the book is not intended as a text for those with little or no relevant background.

Most of the findings detailed in this book and accompanying articles were obtained using psychophysical methods. The psychophysical approach is based on experimental measurements of the dependence of reading performance on physical properties of text such as character size, contrast, color or font. Our principal measure of performance is reading speed. Chapter 2 gives the rationale for choosing reading speed for assessing the role of vision in reading, and compares reading speed to other performance measures.

Theoretical interpretations of results from psychophysical studies enable us to infer the behavior of visual mechanisms in the reading task. Chapter 3 discusses how visual coding of contrast and spatial frequency influences reading performance, and how the structure of the visual field, as represented by the concept of the visual span, has a major impact on reading performance.

A good match between the physical properties of text and the visual capacities of readers, both normal and visually impaired, determines how effectively text can be read. Chapter 4 is devoted to issues of displaying text. How do we measure the legibility of text? What attributes of text enhance or detract from its legibility? Chapter 4 concludes with a summary of guidelines for displaying text.

One outcome of our research on vision and reading has been the development of the MNREAD acuity chart. This chart is intended to provide a standardized assessment of visual function in reading, with applications for both normal and impaired vision. Chapter 5 describes the design principles of the chart and our recommended testing procedures.

Ultimately, reading is a process for converting visual symbols into phonological/linguistic representations (Gelb, 1963). Data acquired visually from text eventually converges on language codes in the brain, likely becoming indistinct from information acquired by speech. Where purely visual processing leaves off and phonological, lexical or other higher-level processing begins is debatable. The line may depend on one's theoretical orientation. Chapter 3 discusses the view that information processing includes an early transformation from sensory representations of letters or words, closely coupled to retinal-image data, to abstract, symbolic representations more suitable for linguistic analysis.

Terms like "stereo vision," "color vision," and "pattern vision" refer specifically to aspects of visual processing that evolved over millions of years to handle especially informative aspects of retinal images. Reading and writing probably originated in about 3000 B.C. in the region now occupied by Iraq with the invention of cuneiform by the Sumerians (although there is still debate among linguists about which cultures may have independently discovered writing systems). Over this period of 5,000 years, the anatomy of human vision has probably remained stable, and it is unlikely that special visual modules have evolved for reading (but see Box 3.2 on the Visual Word Form Area). Presumably, pre-existing visual mechanisms are exploited in reading, but only in recent years have we begun to understand in detail the roles of these mechanisms in reading.

Our research has been strongly motivated by our interest in low vision. Low vision is sometimes defined very broadly as any permanent form of visual impairment affecting everyday function. In our research, we have defined it more specifically as visual impairment resulting in the inability to read the newspaper at a normal reading distance of 40 cm (16 inches) with best optical correction.[1] Many surveys report that difficulty with reading is the primary concern of people entering low-vision clinics (cf. Elliott et al., 1997).

The World Health Organization (2004) estimates that in 2002 there were more than 161 million visually impaired people worldwide, including 124 million with low vision and 37 million who were blind. By a recent estimate, there are about 3.3 million people in the United States over the age of 40 with impaired vision. Of these, about 937,000 are legally blind[2] and another 2.4 million have milder low vision (Eye Diseases Prevalence Research Group, 2004). Among the legally blind, approximately 200,000 are totally blind, that is, have no useful pattern vision. Because the leading causes of visual impairment in the United States are age-related eye diseases—macular degeneration, glaucoma, diabetic retinopathy, and cataract—the prevalence of impaired vision rises steeply with age. There are also racial differences in the causes of impaired vision; age-related macular degeneration is the leading cause among whites (54.4%), and cataract and glaucoma among blacks (> 60%). Because of demographic trends, particularly the aging of the American population, the same study estimates that by the year 2020, these numbers will increase by 70% to 5.7 million people with low vision in the United States, with additional people under 40 years of age not included in these estimates. The vast majority of all people with low vision encounter difficulties with reading.

Our research on low-vision reading has three principal goals—to measure the empirical characteristics of reading deficits in people with impaired vision, to provide an explanation for these deficits, and to translate these findings into useful applications. Chapters 2, 3, and 4 discuss our empirical findings on low-vision reading. We consider theoretical ideas to explain the findings in chapter 3 where we discuss the contrast attenuation model (section 3.3), the visual-span model (section 3.7), and the problems of people with central-field loss (section 3.8). Chapter 4 addresses issues related to displaying text for low vision, including selection of fonts (section 4.2), page navigation and magnifiers (sections 4.4, 4.5, and 4.6), and the effects of color and luminance (section 4.7).

---

[1] According to the 10th revision of the World Health Organization International Statistical Classification of Diseases, Injuries and Causes of Death, low vision is defined as visual acuity of less than 20/60 (metric 6/18), but equal to or better than 20/400 (metric 3/60), or a visual field of less than 20° but equal to or better than 10°, in the better eye with best possible corrections. Blindness is defined as visual acuity of less than 20/400 (metric 3/60), or a visual field of less than 10°, in the better eye with best possible correction. Since there are many people with acuities less than 20/400 who read print visually, given adequate magnification, this dividing line between low vision and blindness seems inappropriate in the context of reading. In our research, we include as low vision those at the low end of the vision scale who are able to read print visually, even if very slowly, given adequate magnification.

[2] Legal blindness—defined in the United States as a corrected visual acuity in the better eye o no more than 20/200, or a visual field of no more than 20°—is used for various legal purposes. The categorization of people as "seeing" or "legally blind" has tended to obscure the many forms of low vision and the functional value of residual vision.

In modern society, reading is one of the most important tasks we do with our eyes, and one of the most impressive in terms of acuity, pattern recognition, oculomotor coordination, and dynamic interaction with higher-level cognition. Recently, Andrew Solomon (2004) wrote in *The New York Times* in praise of reading. He likened the alphabetic code to the genetic code, and emphasized the cultural significance of reading:

> The fact that so much can be expressed through the rearrangement of 26 shapes on a piece of paper—is as exciting as the idea of a complete genetic code made up of four bases: man's work on a par with nature's. Discerning the patterns of those arrangements is the essence of civilization. (p. 17)

Historically, there has been a wide gulf between the literature on basic visual psychophysics and the influence of perceptual factors on real-world tasks such as reading, driving, spatial navigation or social interaction. Although the pioneers who studied vision and reading, such as Javal and Huey (see Box 3.1), emphasized the importance of visual mechanisms, little progress was made for decades. Visual psychophysicists paid little attention to reading and other real-world functions of vision. They focused primarily on characterizing the properties of hypothetical visual mechanisms, and tested hypotheses linking these mechanisms to underlying physiological structures (cf. Brindley, 1970, p. 134). Meanwhile, studies of the psychology of reading focused on cognitive and linguistic constructs such as the nature of lexical access and educational implications for learning to read. Even the vast body of literature on reading eye movements, seemingly a link with visual mechanisms, focused more on implications for higher-level cognitive analysis (see the books by Rayner & Pollatsek, 1989, and Just & Carpenter, 1987). Tinker (1963), in a large body of work including a classic series of 13 articles with D.G. Paterson on "studies of typographical factors influencing speed of reading," used psychophysical methods to study perceptual factors in reading. This work focused almost exclusively on relevance to text legibility, and not the nature of visual processing. Box 1.2 contains citations to Tinker and Paterson's series of 13 articles.

Our interest in low vision and reading brought us face-to-face with the nature of visual-information processing in reading, how visual data are coded by visual mechanisms, and what goes wrong with visual-information processing in eye disease. From the beginning of this research program, one of our major goals has been to link our findings on the psychophysics of reading to contemporary principles of basic visual psychophysics and the structures of the eye and visual system. As a result of our research, reading has become a topic in contemporary vision science, perhaps the archetype for studying vision embedded in a real-world context. Although this book reveals many gaps in our current understanding of vision and reading, and the need for ongoing research, we may dare to hope that progress to date sets the stage for a major accomplishment in the near future—a fully elaborated theory of visual-information processing in reading, linking image formation on the retina to cognition. Achievement of this goal will likely take advantage of joint explorations using psychophysical, cognitive, computational and brain-imaging methods. The future is bright for research on vision and reading.

Reading involves more than visual processing, more than cognition, more than motor control; it requires the integration of all of these processes. Reading is a wonderful example of the flexibility and sophistication of the human brain. In Huey's (1908/1968) words,

> And so to completely analyze what we do when we read would almost be the acme of a psychologist's achievement. For it would be to describe very many of the most intricate workings of the human mind, as well as to unravel the tangled story of the most remarkable specific performance that civilization has learned in all its history. (p. 6)

Box 1.1
Complete Citations for the 20 Articles in the Psychopysics of Reading Series

Note on referencing scheme: In this book, the 20 articles in our series are referenced by number and year. For example, the 15th article (Psychophysics of Reading XV) is referenced like this: (R15, 1996). The complete original citations are included with each article on the accompanying compact disk and are summarized here for convenience. All other citations in the book use the standard American Psychological Association format.

1. Legge, G. E., Pelli, D. G., Rubin, G. S., & Schleske, M. M. (1985). Psychophysics of reading. I. Normal vision. *Vision Research, 25,* 239–252.
2. Legge, G. E., Rubin, G. S., Pelli, D. G., & Schleske, M. M. (1985). Psychophysics of reading. II. Low vision. *Vision Research, 25,* 253–266.
3. Pelli, D. G., Legge, G. E., & Schleske, M. M. (1985). Psychophysics of reading. III. A fiberscope low-vision reading aid. *Investigative Ophthalmology and Visual Science, 26,* 751–763.
4. Legge, G. E., & Rubin, G. S. (1986). Psychophysics of reading. IV. Wavelength effects in normal and low vision. *Journal of the Optical Society of America, A3,* 40–51.
5. Legge, G. E., Rubin, G. S., & Luebker, A. (1987). Psychophysics of reading. V. The role of contrast in normal vision. *Vision Research, 27,* 1165–1171.
6. Rubin, G. S., & Legge, G. E. (1989). Psychophysics of reading. VI. The role of contrast in low vision. *Vision Research, 29,* 79–91.
7. Legge, G. E., Ross, J. A., Maxwell, K. T., & Luebker, A. (1989). Psychophysics of reading. VII. Comprehension in normal and low vision. *Clinical Vision Sciences, 4,* 51–60.

8. Legge, G. E., Ross, J. A., Luebker, A., & LaMay, J. M. (1989). Psychophysics of reading. VIII. The Minnesota low-vision reading test. *Optometry and Vision Science, 66*, 843–853.
9. Parish, D. H., & Legge, G. E. Psychophysics of reading. IX. The stability of eye position in normal and low vision. Unpublished Manuscript 8/89.
10. Akutsu, H., Legge, G. E., Ross, J. A., & Schuebel, K. (1991). Psychophysics of reading. X. Effects of age related changes in vision. *Journal of Gerontology: Psychological Sciences, 46*, 325–331.
11. Legge, G. E., Parish, D. H., Luebker, A., & Wurm, L. H. (1990). Psychophysics of reading. XI. Comparing luminance and color contrast. *Journal of the Optical Society of America, A7*, 2002–2010.
12. Legge, G. E., Ross, J. A., Isenberg, L. M., & LaMay, J. M. (1992). Psychophysics of reading. XII. Clinical predictors of low-vision reading speed. *Investigative Ophthalmology & Visual Science, 33*, 677–687.
13. Ahn, S. J., & Legge, G. E. (1995). Psychophysics of reading. XIII. Predictors of magnifier-aided reading speed in low vision. *Vision Research, 35*, 1931–1938.
14. Beckmann, P. J., & Legge, G. E. (1996). Psychophysics of reading. XIV. The page-navigation problem in using magnifiers. *Vision Research, 36*, 3723–3733.
15. Mansfield, J. S., Legge, G. E., & Bane, M. C. (1996). Psychophysics of reading. XV. Font effects in normal and low vision. *Investigative Ophthalmology & Visual Science, 37*, 1492–1501.
16. Legge, G. E., Ahn, S. J., Klitz, T. S., & Luebker, A. (1997). Psychophysics of reading. XVI. The visual span in normal and low vision. *Vision Research, 37*, 1999–2010.
17. Harland, S., Legge, G. E., & Luebker, A. (1998). Psychophysics of reading. XVII. Low- vision performance with four types of electronically magnified text. *Optometry & Vision Science, 75*, 183–190.
18. Chung, S. T. L., Mansfield, J. S., & Legge, G. E. (1998). Psychophysics of reading. XVIII. The effect of print size on reading speed in normal peripheral vision. *Vision Research, 38*, 2949–2962.
19. Bruggeman, H., & Legge, G. E. (2002). Psychophysics of reading. XIX. Hypertext search and retrieval with low vision. *Proceedings of the IEEE, 90*, 94–103.
20. Legge, G. E., Mansfield, J. S., & Chung, S. T. L. (2001). Psychophysics of reading. XX. Linking letter recognition to reading speed in central and peripheral vision. *Vision Research, 41*, 725–34.

Box 1.2
Citations to M. A. Tinker's Classic Series of 13 Articles

Miles A. Tinker 1893–1977 published a large and influential body of research on reading and perception. The centerpiece of his work was a series of 13 articles with D.G. Paterson published under the general heading, "Studies of Typographical Factors Influencing Speed of Reading." Citations to these 13 articles are given here. Tinker's (1963) influential book, *The Legibility of Print,* summarized the literature through the mid-20th century.

i. Tinker, M. A., & Paterson, D. G. (1928). Influence of type form on speed of reading. *Journal of Applied Psychology, 12,* 359–368.

ii. Paterson, D. G., & Tinker, M. A. (1929). Size of type. *Journal of Applied Psychology, 13,* 120–130.

iii. Tinker, M. A., & Paterson, D. G. (1929). Length of Line. *Journal of Applied Psychology, 13,* 205–219.

iv. Paterson, D. G., & Tinker, M. A. (1930). Effect of practice on equivalence of test forms. *Journal of Applied Psychology, 14,* 211–217.

v. Tinker, M. A., & Paterson, D. G. (1931). Simultaneous variation of type size and line length. *Journal of Applied Psychology, 15,* 72–78.

vi. Paterson, D. G., & Tinker, M. A. (1931). Black type versus white type. *Journal of Applied Psychology. 15,* 241–247.

vii. Tinker, M. A., & Paterson, D. G. (1931). Variations in color of print and background. *Journal of Applied Psychology, 15,* 471–479.

viii. Paterson, D. G., & Tinker, M. A. (1932). Space between lines or leading. *Journal of Applied Psychology, 16,* 388–397.

ix. Tinker, M. A., & Paterson, D. G. (1932). Reduction in size of newspaper print. *Journal of Applied Psychology, 16,* 525–531.

x. Paterson, D. G., & Tinker, M. A. (1932). Style of type face. *Journal of Applied Psychology. 16,* 605–613.

xi. Tinker, M. A., & Paterson, D. G. (1935). Role of set in typographical studies. *Journal of Applied Psychology, 19,* 647–651.

xii. Paterson, D. G., & Tinker, M. A. (1936). Printing surface. *Journal of Applied Psychology, 20,* 128–131.

xiii. Tinker, M. A., & Paterson, D. G. (1936). Methodological considerations. *Journal of Applied Psychology, 20,* 132–145.

2 Measuring Reading Speed

Gordon E. Legge

Our focus in this book is on the role of vision in reading. Reading speed, sometimes called "reading rate," has been our primary psychophysical measure. This chapter gives the rationale for choosing reading speed. Section 2.1 describes three methods for measuring reading speed. Section 2.2 discusses some testing subtleties and their effects on reading speed. Section 2.3 asks how quickly people read, and briefly reviews findings for different groups of readers and sensory modalities (vision, touch and hearing). Section 2.4 compares speed to other measures of reading performance.

2.1 THREE METHODS FOR MEASURING READING SPEED

Reading performance has been measured in many ways. In the eye clinic, reading acuity, representing the smallest print that can be read, has been the dominant measure. In education, reading comprehension is the principal measure. *Reading speed* in words per minute (wpm) has been used to study both educational and perceptual aspects of reading (Abell, 1894; Carver, 1990; Huey, 1908/1968; Tinker, 1963;). We adopted reading speed as our primary measure. At the beginning of our psychophysical studies (R1, 1985; R2, 1985). We developed several methods for measuring reading speed that have proven to be reproducible and sensitive to visual factors. Reading speed is a functionally significant measure, especially for people with low vision who often read slowly. In section 2.4, we compare reading speed to some alternative measures of reading performance.

Reading speed captures two essential properties of the visual part of reading: recognition of written words, and the processing of a rapid temporal sequence of stimuli. In chapter 3, we argue that these two key processes are constrained by early sensory mechanisms within the visual system.

Drifting-Text Method

We can use psychophysical threshold methods to measure reading speed by pushing participants to their fastest performance. We began by developing a *drifting-text* technique (originally termed "scanned" text; R1, 1985). A single line of text drifts from right to left across a display screen at a constant rate, similar to overlay weather alerts on commercial TV. Critical to the procedure, the participant reads the text aloud (oral reading method) so that the experimenter can objectively score accuracy on a word-by-word basis. Faster and faster drift rates are tested In a sequence of trials to find a threshold value, that is, the fastest drift rate yielding accurate reading performance. A plot of the percentage of words read correctly as a function of drift rate is called a psychometric function, and can be used to determine reading speed.

Figure 2.1(a) shows a sample psychometric function for one participant. Typically, for a range of slow drift rates, the participant reads the drifting text perfectly (100% of words identified accurately). Eventually, a critical drift rate is reached at which there is a very rapid transition from perfect reading to inaccurate reading. For the example in panel (a) the critical drift rate is 145 words per minute (wpm). Accuracy decreases very rapidly at higher drift rates. In general, the critical drift rate depends on stimulus properties such as character size, and also the visual status of the participant.

Figure 2.1(b) shows the transformation of the reading accuracy data in panel (a) to reading speed. *Reading speed* is computed as the drift rate in wpm times the proportion of words read accurately. For instance, if 100 wpm appear on the screen at the right margin, and the participant reads 90% of them correctly, the corresponding reading speed is 90 wpm. Because of the sharp transition from accurate to inaccurate reading at the critical drift rate, there is a well-defined *maximum reading speed*, as shown by the peaked curve in panel (b). The maximum reading speed occurs within a very narrow range of drift rates at or near the critical value. Psychophysical threshold methods can be used to identify the critical drift rate. Although staircase methods have sometimes been used, we frequently used a method of limits; the experimenter varied the drift rate on a trial-by-trial basis to find the participant's critical value, and then computed the maximum reading speed from the corresponding drift rate and proportion of words read correctly.

Notice that this method takes into account both the rate of stimulus presentation and the accuracy with which words are read. By identifying the drift rate yielding maximum reading speed, we avoid problems with speed-accuracy trade-offs. Carver (1990) described a related measure termed *reading efficiency*. For his measure, the number of words read silently in a fixed period of time is used to compute *reading rate*. Then, multiple-choice comprehension questions are administered.

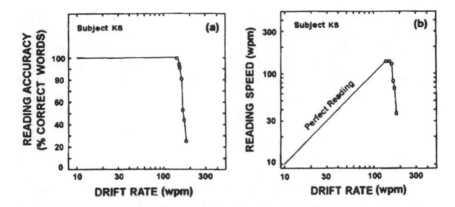

Figure 2.1. (a) Reading accuracy as a function of drift rate. Percentage of words correctly read is plotted as a function of drift rate for observer K.S. The horizontal solid line at 100% correct represents perfect reading. (b) The data of panel (a) have been replotted as reading speed vs drift rate. Reading speed is the product of drift rate and proportion correct. The diagonal solid line represents perfect reading. The data in this example are taken from an experiment on matrix sampling (see R1, 1985, Fig. 4). The effects of sampling are discussed in section 4.3.

Carver's (1990) reading efficiency is the product of reading rate and percent correct comprehension. Carver's (1990) measure is intended for educational applications where reading comprehension is the primary objective. Comprehension is less tightly coupled to vision than reading accuracy (see section 2.4 and R7, 1989), so our reading-speed measure is more appropriate for assessing the impact of visual factors.

One advantage of the drifting-text method, pertinent to low vision, is that very large characters can be displayed. As we discuss in section 4.5, only about five characters need be visible at a time to achieve maximum reading speeds with drifting text (R1, 1985; R2, 1985; R14, 1996). For example, suppose a screen, subtending 60° (easy to achieve with a video monitor and short viewing distance), is used to display drifting text. If five characters fill the horizontal extent of the screen, the characters will be very large, each subtending about 12° of visual angle. The drifting-text method permitted us to study the dependence of reading speed on character size over a wide range (see section 3.2).

One concern with the drifting-text method is that it differs from everyday reading in the pattern of eye movements. Eye movements in static text consist of a series of fixations, typically lasting 200 to 250 ms, separated by saccadic eye movements averaging about 7 characters in length. With drifting text, the eyes typically land on a word near the right margin of the screen, track it as it drifts leftward across the screen (smooth-pursuit eye movements) and then saccade back to the right margin to pick up the next word. The cycle repeats, and a saw-tooth pattern[1] similar to optokinetic nystagmus results (R1, 1985; Buettner, Krischer, & Meissen, 1985). The resulting sequence of retinal images mimics the sequence for static text; there is a series of foveal pauses on words, separated by saccades spanning a few letter spaces. The functional similarity of eye-movement patterns implies that the spatiotemporal characteristics of retinal images are very similar for reading drifting and static text. We have shown that the dependence of reading speed on stimulus variables such as color and luminance (R4, 1986) and contrast (R5, 1985) is the same for drifting and static text.

Historically, we first implemented drifting text by using a mirror-galvanometer to scan the text through the field of view of a video camera. The camera was connected to a closed-circuit TV for stimulus display. Then we upgraded to a system in which Text, printed on cards, was placed flat on a movable platform beneath the TV camera. A motor, pulleys, and rubber belts pulled the platform along a track so that the text swept beneath the camera's field of view. Speed was controlled by adjusting a potentiometer that regulated the voltage driving the motor. Eventually, we graduated to a computer-based video frame buffer (Grinnell GMR274) with pan control (R5, 1987).

Current commercially available computer screen-magnification programs for low vision, such as ZoomText (AI-SQUARED, Manchester Center, Vermont, http://www.aisquared.com), include drifting-text options with adjustable speeds for reading.

Rapid Serial Visual Presentation (RSVP) Method

In the RSVP method, individual words are presented sequentially at the same location[2] on a display screen. The RSVP rate is controlled by adjusting the exposure time for each word. RSVP reduces or eliminates the role of eye movements in reading. RSVP was originally used in cognitive studies of word recognition in reading (Forster, 1970), and was introduced into psychophysical studies of normal and low-vision by Rubin and Turano (1992, 1994). This technique lifts a ceiling on normal reading speed imposed by the latency for eye movements; RSVP reading speeds are typically much higher than speeds for static text. For example, Rubin and Turano (1992) reported an average reading speed of 1,171 wpm for RSVP text compared with 303 wpm for static text.

---

[1] A similar pattern of reading eye movements—periods of smooth-pursuit tracking separated by saccades—occurs when normally sighted subjects read with hand-held or spectacle-mounted optical magnifiers (A. R. Bowers, 2000). Imagine, for example, keeping your head steady while reading a letter with a spectacle-mounted magnifier. You would move the text through the field of view of the magnifier, producing a moving image similar to drifting text on a display screen.

[2] In creating RSVP displays, one must decide whether a fixed reference point on the screen is used for left-justifying words or for centering words. We have typically left-justified the words, reasoning that subjects will do better if they have precise knowledge about the locations of the leading letters of words.

The RSVP method shares two of the advantages of drifting text: it permits the display of very large characters on the screen, and it leaves the rate of stimulus presentation in the hands of the experimenter. Moreover, the reduced dependence on eye movements helps decouple visual and motor factors in reading, resulting in performance that is more directly influenced by visual factors.[3] From a technical point of view, RSVP text is easier to produce on a computer screen than drifting text.

Because of its advantages, in several of the later studies in our series, we have used the RSVP method (R16, 1997; R18, 1998; R20, 2001). Rather than use a method of limits (as described above for drifting text) to locate RSVP exposure times yielding maximum reading speeds, we have preferred to use a method of constant stimuli to obtain psychometric functions of reading accuracy as a function of RSVP exposure time. For instance, in R18 (1998), we measured RSVP reading speeds for several retinal eccentricities and print sizes. Each trial consisted of the presentation of one sentence with a designated exposure time for each word. The participant read the sentence aloud and the experimenter recorded the number of errors. For each condition, there were six RSVP trials at six different exposure times spanning about a factor of 10. The proportion of words read correctly was computed for each exposure time, based on the six trials, resulting in a psychometric function of percent correct versus exposure time. Examples can be seen in Figure 2 of R18. Reading speed was based on the exposure time yielding 80% correct on the fitted curve. For example, if an exposure time of 200 msec per word yielded 80% correct, the reading rate was 5 words per second equals 300 wpm.

The RSVP psychometric functions have a nice property for psychophysics. They are quite steep, having a rapid transition from inaccurate to accurate reading for a small change in exposure time. Except for print sizes near the acuity limit (where the psychometric functions are shallower), normally sighted participants require a roughly constant factor of two in exposure time for improvement from 50% to 84% accuracy (R18, 1998). Preliminary findings (Chung & Legge, 1997) indicate that a similar rule holds for low-vision participants.

Flashcard Method

Standardized methods for measuring reading speed in education typically use printed text on paper, which we term "static text" to be distinguished from drifting or RSVP text. For example, in the Nelson–Denny reading test, a participant reads several passages (each about 600 words) and answers multiple-choice questions following each passage. The results are used to compute a comprehension score. On the first passage, the examiner interrupts reading after 1 min, and the participant indicates the point he or she has reached in the passage. This determines the number of words read, yielding a rate score (see Carver, 1990, chap. 13 for more

---

[3]RSVP is particularly useful for measuring reading speed in normal peripheral vision (R18, 1998). This is because normally sighted subjects, who fixate foveally, have difficulty in making reading saccades appropriate for text viewed eccentrically. RSVP reduces the need for such saccades.

discussion of this test and associated artifacts). This style of test has the advantage of measuring reading in a fairly natural context. A major disadvantage is that it leaves the participant's strategy, including skimming or skipping, uncontrolled and largely undetected.

In 1989, we introduced a test to help bridge the gap from the drifting-text method to static text (R8, 1989). Our motivation was twofold: (a) to provide a standardized, computer-based test of reading speed, sensitive to visual factors and suitable for both research and clinical use; and (b) to use static text. The test was called the Minnesota Low-Vision Reading Test, or MNREAD[4] for short. The test involved presentation of a "flashcard" on the computer screen, formatted into four lines of 13 characters each, in a fixed-width font. There were four subtests—continuous sentences or unrelated words, and black letters on a white background or white letters on a black background. Examples are shown in Figure 1 of R8. The flashcard fills most of the screen so that at a near viewing distance, the letters subtend a large visual angle, suitable for testing low vision. For standardization, we recommended a viewing distance at which the letters subtended 6°, but by varying viewing distance, any angular character size could be achieved.

In the test, a flashcard is presented for a specified exposure time. The participant is asked to read it aloud as quickly and accurately as possible. If the entire flashcard is read correctly, a shorter exposure time is used on the next trial. The exposure times are adjusted to find a value for which the participant reads more than half the words but is unable to complete the entire flashcard accurately. The experimenter counts the number of words read accurately and computes a corresponding reading speed. For example, if the exposure time is 2 sec and the participant reads 8 words correctly, the corresponding reading speed is 4 words per second, equal to 240 wpm. The test is relatively quick and easy to administer, and has high test–retest correlation, $R = 0.88$ (R8, 1989).

This method couples the advantages of oral reading and timed exposures from the drifting-text and RSVP methods with the use of static text. Although the text lines are short (13 characters), performance requires reading eye movements including forward saccades and return sweeps. The use of a block of text is more similar to everyday reading than a single drifting line or an RSVP sequence of single words.

How does reading speed compare for the drifting-text and flash-card methods? In a recent study, A. R. Bowers, Woods, and Peli (2004) found no significant differences in maximum oral reading speeds for drifting text (termed "horizontal scrolling" in their study) and static text (termed "page") for either normal or low vision. We did find modest differences in a direct comparison (R8, 1989). Normal participants read static text faster than drifting text, with the difference depending on character size (R8, 1989; Fig. 3). Unlike normal participants, a heterogeneous group of 27 low-vision participants read drifting text slightly faster (average of 15%) than static text, although the speeds were highly correlated at $R = 0.88$ (R8,

---

[4]Later, the term MNread changed to MNREAD, and was applied to a reading-acuity chart in which reading speed is measured as a function of print size. This chart is discussed in Chapter 5.

1989). Why might people with low vision read drifting text faster than static text? Many people with low vision have unstable fixation. Drifting text may help by entraining eye movements, or by eliminating the need to make return eye movements from the end of one line to find the beginning of the next line.

Subsequently, we created a version of the flash-card test on printed cards (Ahn, Legge, & Luebker, 1995). The purpose of converting the test from the computer to hardcopy was to simplify it for clinical use. Instead of presenting the flashcards for timed periods (computer method), the participant is shown the flashcard and asked to read it as quickly and accurately as possible. The time is recorded on a stopwatch and the number of errors recorded (words left out or misread). Reading speed is estimated as the number of words read correctly divided by the reading time. We learned from this study that an estimate of reading speed from a single printed flashcard (containing only 56 characters) is fairly accurate (standard deviation is 18% of the mean). Later, we incorporated this method of measuring reading speed into the protocol for the MNREAD Acuity Chart (see chap. 5.)

Clinical Potential of the Flashcard Method

In R13 (1995), we established the validity of the flashcard test for assessing low-vision reading performance. Forty low-vision participants read passages of printed text on paper with their preferred magnifiers. Reading speed was estimated from the time taken to complete the passage. Reading speeds measured with the computer-based flashcard method (R8, 1989) were highly correlated with the magnifier-aided reading speeds, accounting for 79.7% of the variance. This finding revealed a potential use of the flashcard method for rehabilitation. It provides an assessment of the impact of visual factors on reading that is predictive of real-world reading with a magnifier.

Why not use traditional clinical measures of vision to predict reading performance rather than a new test of reading speed? There is growing agreement among clinical researchers that evaluation of eye therapies or rehabilitation strategies should include an assessment of the impact on quality of life and the ability to perform important everyday tasks such as reading and mobility. It would be convenient if standard clinical tests of vision, such as letter acuity, were predictive of performance on everyday tasks. We could then measure acuity, or some other clinical measure, before and after the intervention as a surrogate for real-world function. Unfortunately, the correlations between clinical vision measures and real-world performance are usually weak, as in driving (Owsley & McGwin, 1999) and reading. In R12 (1992), we asked if a set of simple clinical measures, including letter acuity, could be used to predict reading speed. One hundred and forty one patients who entered the low-vision clinic of the Minneapolis Society for the Blind received thorough eye examinations and a test of reading speed. Letter acuity accounted for only 10% of the variance in reading speeds, similar to the weak correlations obtained in studies with smaller participant samples (R2, 1985; R13, 1995). The lesson appears to be that rather than relying on predictions from clini-

cal tests of visual function, clinical assessment of low-vision reading may be accomplished most easily with a suitably designed reading test. Chapter 5 describes
the MNREAD acuity chart, a reading test that has grown out of our psychophysical
research.

2.2 SUBTLETIES IN THE MEASUREMENT OF READING SPEED

Several procedural and scoring factors complicate psychophysical measurements
of reading speed, and frustrate quantitative comparison across studies. Some of
these problems were discussed in the previous section. In this section, we briefly
treat four additional topics: text difficulty, instructions to the participant, oral versus silent reading, and context effects.

Text Difficulty, Word Length, and Carver's Metric

Carver (1976, 1990) pointed out that measuring reading speed in wpm can be affected by mean word length. In general, easier texts (lower grade level) have
shorter words. Carver (1976, 1990) presented data demonstrating that reading
speeds are nearly invariant across text difficulty provided 1) speed is measured in
characters per unit time rather than words per unit time, and 2) the grade level of
the text is below the participant's reading level. Carver (1976, 1990) defined a
"standard-length word" to be six characters, so reading speed in characters per unit
time can be equivalently expressed in units of standard-length words per unit time.
For instance, the current sentence takes 78 keystrokes, including punctuation. It
has $78/6 = 13$ standard-length words, and 10 actual words (where the 2-digit number is counted as a single word). If the sentence were read perfectly in 2 sec,
Carver's (1976, 1990) metric yields a rate of 39 characters per second, equivalent
to 6.5 standard-length words per second or 390 standard-length wpm (slwpm).
The corresponding conventional reading speed (based on 10 real words in the sentence) would be 5 words per second or 300 wpm.

Carver's (1976, 1990) character-based metric reduces variability in reading
speed measurements due to differences in mean word length. For this reason, we
have use the Carver (1976, 1990) metric in several of the later papers in our series.
We have also followed recommendations by Carver (1990), and Whittaker and
Lovie-Kitchin (1993) to use text that is several grade levels below the participant's
reading level, as a way of minimizing the impact of text difficulty on reading
speed.

Instructions to Participants

Carver (1990) showed that instructions to participants can have a major impact on
their reading speed. Instructions to learn or memorize details will result in much
slower reading than instructions to skim for gist or search for key words.

Instructions such as "read at your ordinary, comfortable pace" will yield slower reading speeds than instructions to read at maximum speed. Huey (1908/1968) found a nearly 50% difference in silent reading speeds for a group of graduate students given these two types of instructions (mean of 337 wpm for ordinary reading and 492 wpm for maximum speed). Why would people ordinarily choose to read at sub-maximum speeds? As discussed below in section 2.3, and also in R7 (1989), comprehension declines as one approaches maximum reading speed. A person's everyday reading speed implicitly resolves a speed-accuracy tradeoff between maximizing accuracy of comprehension and minimizing the time to read text.

The three methods described in section 2.1 require participants to read all words in sequence under time pressure, presumably challenging the limits of visual processing. The use of oral reading allows the experimenter to deduct credit for words skimmed or misread. When static text is used, as in the flashcard method, the instructions urge participants to read at their visual limits, "Read as quickly and accurately as you can, keeping errors to a minimum ...."

Keep in mind that for some research questions, we may want participants to read at their ordinary pace rather than trying to maximize their speed. For example, in R17 (1998), our goal was to compare low-vision reading speeds for several types of display devices. We asked participants to silently read passages of about 70 words in length. The participants were instructed to read the passages as they would normally read a magazine article but without skipping or skimming. They were also told that they would be asked multiple-choice questions about the main ideas in the passages.

Oral and Silent Reading of Short and Long Passages

Oral reading is convenient for objective psychophysical measurement, because the experimenter can score the participant's accuracy in reading words. But does oral reading produce speeds similar to the more common silent reading, when both are tested with similar instructions? Empirical findings indicate that the answer is yes for the very short passages used in the drifting-text method (R1, 1985; R4, 1986; R5, 1987) and the computer-based flashcard method (R8, 1989).

One seeming concern with oral reading is that the participant's voice may lag behind visual decoding. This is not a problem when the text passages are short enough to be retained in short-term memory and the text-presentation rate is controlled by the experimenter, as in the drifting, RSVP and computer-based flashcard methods. In these cases, the participant's verbal response is used only to score accuracy, and can safely finish after the termination of the text stimulus. In some cases, such as RSVP sequences, display of the stimulus sentence may even be complete before the participant begins to speak. For example, normally sighted participants may read perfectly a 6-word RSVP sentence, presented at 100 msec per word. The entire sentence requires only 600 msec for display, and may be complete before the participant utters the first word.

The printed-cards version of the flashcard method (see section 2.1) may have a ceiling imposed by the participant's speaking rate. This is because timing is not independently controlled by the experimenter, but ends when the participant finishes talking. Fast readers may complete the visual processing of text ahead of their voice. Most people can easily speak at rates of at least 200 wpm. For most low-vision participants, speeds are well below 200 wpm, so rate of speaking is unlikely to be a limiting factor for most of them.

In oral reading, eye-movement measurements show that the reader's voice usually trails the eyes through the text. The voice can lag behind the eyes by at least a couple of words (the "eye-voice span"), and the eyes sometimes stall for time with long fixations, or even make regressive saccades to let the voice catch up (Rayner & Pollatsek, 1989, p. 181). When the reader's maximum speaking rate is slower than the rate of visual decoding of text, the speaking rate will impose a ceiling on oral reading speed for lengthy passages. A vocalization ceiling explains why estimates of oral reading speeds are typically found to be slower than silent reading for lengthy passages of text. For adults, it is commonly found that silent rates are about 50% faster than oral rates (Bouma & de Voogd, 1974; Huey, 1908/1968).

Tests of silent reading speed are favored in education because they are easy to administer and natural for students. Carver (1989) showed that children's silent rates are only about 12% faster than their oral reading rates.

It might also be argued that oral reading invokes phonological analysis that makes it qualitatively different from silent reading. Since the early work of Huey (1908/1968), it has been argued that "silent speech" can accompany silent reading. Carver (1990, chap. 4) has reviewed literature indicating that phonological analysis ("silent speech") is a normal and nonharmful accompaniment to most silent reading. This being the case, the major difference between silent and oral reading is the need for articulation in the latter.

The following summary points from the above two subsections are relevant to psychophysical measurements of reading speed:

- For normally sighted people, maximum silent reading speed is usually about 50% faster than ordinary or "natural" silent reading speed, even when participants are not deliberately skimming or skipping.
- For long passages of text, the silent reading speeds of normally sighted participants are usually faster than oral reading speeds (typically about 50% faster for adults), because of the ceiling imposed by articulation for oral reading.
- For procedures in which maximum reading speed is computed from the display time of short texts, and oral reading is used only to check for accuracy, oral and silent reading speeds are approximately the same.

Context Effects

Some researchers who study vision and reading prefer to use unrelated strings of words to minimize nonvisual, top-down influences. Others prefer the greater ecological validity of continuous sentences. Language contains syntactic and semantic relationships between words that make continuous text more predictable than random words. We might expect this increased predictability to result in faster reading of sentences than random sequences of words. Legge, Hooven, Klitz, Mansfield, and Tjan (2002) analyzed the effect of this sort of predictability on the performance of an ideal-observer model of reading (Mr. Chips). The model exhibited about a 30% increase in reading speed associated with context. The Mr. Chips model interprets the improvement due to context in terms of an increase in mean saccade length associated with the greater predictability (lower entropy) of words in sentences. Bullimore and Bailey (1995) found a similar context advantage for normal readers. Their study used static text presentation. Stronger context effects in normal reading have been reported for dynamic forms of text presentation: RSVP reading (E. M. Fine & Peli, 1996; R18, 1998) and drifting text (E. M. Fine & Peli, 1996).

Should we expect low-vision reading speed to show the same benefit from context as normal reading speed? If people with low vision have an extra cognitive load associated with decoding degraded visual input, they may have fewer cognitive resources available for contextual analysis, resulting in weaker context effects. In support of this view, some studies have shown that people with normal vision show a reduced context effect under challenging viewing conditions, such as reading with peripheral vision (R18, 1998; Latham & Whitaker, 1996), or with simulated cataracts (E. M. Fine, Peli & Reeves, 1997). (However, E. M. Fine, Hazel, Petre, & Rubin, 1999, found no decrease in the benefit of context in normal peripheral vision and a greater benefit from context in the left peripheral field compared with central vision or the lower visual field.)

On the other hand, larger context effects in low vision might be expected from studies of normally sighted participants using a sentence-priming paradigm (cf. Stanovich, 1980; Stanovich & West, 1983). In the sentence-priming paradigm, the participant reads a sentence aloud with the last word ("target" word) missing. Then the target is presented, and the participant names it as quickly as possible. Target recognition times are compared for conditions varying in the predictability of the target given the context. A robust general finding is that overall context effects are smaller in good readers than poor readers (for a review, see Stanovich, 2000). Stanovich and West (1983) also showed that by degrading the visual quality of the target word, good readers could be made to behave more like poor readers in showing amplified context effects. The pattern of results implies a greater context effect (top-down effect) when visual input is deficient.

The empirical results on the size of context effects in low vision are mixed. In R8 (1989, Fig. 2), we showed a 15 to 30% reading-speed advantage for sentences over random words in a heterogeneous group of 147 low-vision participants, roughly consistent with findings from normal readers. The correlation between speeds measured with the two types of text was very high, $R = 0.95$. The high correlation implies that very similar information is obtained from the reading-speed measurements with sentences or random strings. E. M. Fine and Peli (1996) found larger context effects overall, about a two-fold increase in reading speed due to context, but no significant difference between normal participants and a group of low-vision participants with central-field loss. Bullimore and Bailey (1995) found a substantially larger context effect for low-vision readers with age-related macular degeneration (AMD); the AMD participants read continuous text more than twice as fast as text composed of random words. Finally, in a recent study, Sass, Legge and Lee (2006) directly compared context effects in normal and low-vision participants. They found larger context effects for the normal participants. The group of 20 low-vision participants, many of whom had AMD, averaged a 51% speed advantage for continuous text over scrambled text.[5] The group of 20 normally sighted participants had a larger context advantage of 85%. Across all normal and low-vision participants in the Sass et al. study, there was a moderate correlation between the size of the context effect and absolute reading speed; the slower the reader, the smaller the percent increase in reading speed due to context.

To summarize, maximum reading speeds for both normal and low-vision participants benefit from sentence context.[6] The magnitude of context effects, and how they compare between normal and low vision, differ across studies.

For psychophysical measurement of reading speed as a function of some stimulus parameter, it may not matter whether sentences or random strings are used because the results are highly correlated.

2.3 HOW FAST DO PEOPLE READ?

Given the methodological factors discussed above, it is not surprising that substantial variations in reading speed are reported across studies. Added to the influence of methods are wide individual differences, and even the modality of reading—vision, touch or hearing. In this section, we briefly review some of the variations due to individual factors, and sensory modality.

Reading Speeds of Normally Sighted People

Taylor (1965) summarized normative data on reading performance of 12,359 students from first grade to college level. Mean reading speed for the college students was 280 wpm. This is consistent with the typical range of 250 to 300 wpm often given as a rule of thumb.

In our psychophysical studies, we tested small groups of normal participants, often across a wide range of a stimulus parameter such as character size. Table 2.1 lists mean reading speeds in wpm from four of our studies for normally sighted partici-

---

[5]The Sass, Legge, and Lee (2006) study also contained a comparison of reading speeds for spaced and unspaced text. The percentage improvements in reading speed due to context cited in this chapter are based on results combined across spaced and unspaced text conditions.

[6]A similar context effect is found for Braille reading (Legge, Madison, & Mansfield, 1999). Continuous text is read faster than scrambled sentences: 31% faster for uncontracted Braille, and 40% faster for contracted Braille.

TABLE 2.1

**Mean Reading Speeds and Standard Deviations
for Normally Sighted Young Adults**

	Method				
	Drift	Flashcard	Flashcard	RSVP	Rate Level Test[e]
Character Size (°)	0.5	1.0	0.5	> CPS[c]	Not controlled
Mean (wpm)	352	385	510	862	
Mean (slwpm)[a]	370	306	413	690	290
% S.D.	12	21	35	21	17
N	16	4	10	6	43
Source	R10 (1991)	R11 (1990)[b]	Sass et al. (2006)	R18 (1998)[d]	Carver (2000)

Drift, flashcard, and RSVP methods were used to measure maximum oral reading speed. The Rate Level Test measured ordinary silent reading speed.

[a] slwpm: standard-length words per minute (see text, and Carver, 1990) for definition; [b] 95% luminance contrast condition; [c] Reading speed based on print sizes exceeding *critical print size* (CPS), as discussed in Section 3.2; [d] Central-vision condition; [e] Test of "typical reading speed" using simple text. Data averaged for forms A and B from 43 graduate students.

pants. A separate row converts the mean reading speeds to standard-length words per minute (slwpm), Carver's (1976, 1990) recommended metric.[7] Standard deviations in the table are expressed as a percentage of the mean reading speed. The last row of Table 2.1 includes values from the Rate Level Test (Carver, 2000). This test provides measures of typical reading speed for simple text printed on paper.

Even after conversion to slwpm, the mean speeds from our psychophysical studies are higher than Taylor's (1965) cited value of 280 wpm for college students, probably reflecting our methods for pushing participants to their maximum reading speeds. Particularly noteworthy, the mean RSVP reading speed for six young normally sighted participants (R18, 1998) was 862 wpm. RSVP speeds are typically much higher than those for static (or drifting) text because of the reduced need for eye movements. In R20 (2001), we argue that RSVP reading speeds in central vision are fundamentally limited by spatiotemporal limitations on letter recognition. This argument is reviewed in section 3.7.

---

[7] For instance, in Sass, Legge, and Lee (2006), all the test sentences were 56 characters in length and contained an average of 11.5 words. A sentence of 56 characters contains 9.33 standard-length words (one standard length word is six characters). The mean value of 510 wpm in Table 2.1 is based on scoring of the actual number of words read correctly. Conversion to speed in standard-length words per minute (slwpm) involves reducing the mean speed by the ratio 9.33/11.5 = /81 yielding a speed of 413 slwpm.

The mean reading speeds cited in Table 1 all come from groups of normally sighted young adults. They were all well-educated, native-English speaking readers. As such, the values are representative of maximum reading speeds for a fairly homogeneous population of good readers. Nevertheless, there is considerable variability, with standard deviations ranging from 12% to 35% of the mean.

The wide range of reading speeds (and other measures of reading performance) among normally sighted children and adults is a topic of fundamental importance for educational researchers. Before considering perceptual or cognitive influences, it should be recognized that poor reading, even among normally sighted people, may have a variety of causes. In the United States, low literacy is a major societal problem. According to the National Institute for Literacy (http:// novel.nifl.gov/nifl/faqs.html), the 1992 National Adult Literacy Survey revealed that between 21% and 23% of the adult population had level 1 literacy skills (i.e., they were unable to fill out most forms or read a simple story to a child). Other nonvisual causes of poor reading include reading in a nonnative language, cognitive or neurological dysfunction, and dyslexia.

What explains variations in reading speed among well-educated, normally sighted adults? Jackson and McClelland (1975, 1979) investigated factors that distinguish fast from slow reading among normally sighted college students. They asked whether differences in spatial or temporal visual thresholds could account for this variation. They concluded that sensory thresholds were unlikely to be the cause because groups of "fast" and "normal" readers did not differ significantly in threshold exposure times for identifying isolated letters, or in the accuracy for recognizing letters presented at various distances eccentric to fixation. Jackson and McClelland's (1975, 1979) findings do not rule out sensory constraints on reading speed. It is possible for example, that visual factors common to all normally sighted readers impose overall bounds on reading performance, while individual variations are controlled by individual differences in higher-level cognitive or linguistic factors. It is also possible that Jackson and McClelland (1975, 1979) did not measure the right visual thresholds. More recent studies have found correlations between performance on word-recognition tasks and thresholds for detecting visual motion coherence[8] in unselected samples of children and adults (Cornelissen, Hansen, Gilchrist, et al., 1998; Cornelissen, Hansen, Hutton, et al., 1998; Talcott et al. 2002). It is plausible that performance on visual motion-discrimination tasks taps underlying variations in speed or accuracy of temporal coding relevant to reading. Loosely consistent with these findings on motion perception, Jackson and McClelland (1979) found that their "fast" readers had shorter reaction times on a letter-matching test. They interpreted this advantage, not as a difference in sensory capacity, but as a difference in accessing speed for memory representations of letters or words. Similarly, Stanovich (1980) has sum-

---

[8]In a motion coherence task, subjects view a display composed of randomly moving dots. "Coherence" is introduced by requiring that a fraction of the dots move in the same direction. A threshold fraction can be determined for which subjects are able to detect the coherence.

marized findings showing that recognition time for isolated words is highly correlated with individual differences in reading fluency. We return to the connections between letter recognition, word recognition and reading speed in chapter 3.

The issue of individual differences in reading performance is intertwined with the definition of dyslexia. Developmental dyslexia[9] has been defined in DSM IV as decreased reading ability relative to intelligence in the absence of neurologic disorder, sensory impairment or inadequate schooling (American Psychiatric Association, 1994). The premise of this definition is that some unique etiology is associated with poor reading in people of average or higher intelligence. This premise has been challenged on both ethical and scientific grounds (cf. Stanovich, 1999; Stanovich, 2000, chap. 17). In particular, there appears to be no compelling scientific argument that the word-decoding problems underlying reading disability are qualitatively different for poor readers with high intelligence than for other poor readers. Moreover, there is even controversy about whether dyslexics form an etiologically distinct subgroup or whether they constitute the tail of a distribution of reading skills that includes "normal" readers. Shaywitz, Escobar, Shaywitz, Fletcher, and Makuch (1992) concluded from a large study of children that those categorized as reading disabled lie on the tail of a normal distribution of readers. They make an analogy between dyslexia and high blood pressure (hypertension); neither is an all-or-none phenomenon. Adopting this perspective, "dyslexia" is a term that refers to especially poor reading performance along a continuum, and the key issue is what perceptual or cognitive factors underlie the variation along the continuum.

The current consensus is that the core deficit in developmental dyslexia is phonological (Habib, 2000; Shaywitz, 1998). Afflicted children have trouble mastering and using the relationships between printed letters or groups of letters and their sounds. But many studies have also found visual factors that differentiate dyslexic from normal readers, including sensitivity to contrast and flicker. An empirical link between motion perception and reading ability provides additional support for the magnocellular theory of dyslexia. Box 2.1 discusses this theory.

Speed Reading

A skimming strategy accounts for the very high reading speeds (3 to 10 times normal rates) associated with "speed reading." For relevant data and reviews, see Just and Carpenter (1987, chap. 14), and Carver (1990, chap. 19). These authors have shown that speed reading typically amounts to sparse sampling of text, based on strategic placement of eye fixations. There is no compelling evidence that speed readers have enhanced perceptual abilities permitting them to encode extra information on each fixation. What may be learned in courses on speed reading is an effective strategy for skimming through easy text.

---

[9]The terms "dyslexia" and "reading disability" are considered here as interchangeable. SDM-IV refers to the condition as "Reading Disorder," but notes that it is also referred to as "dyslexia."

Box 2.1
Role of the Magnocellular Pathway in Reading and Dyslexia

Since about 1980, the prevailing hypothesis for a visual deficit in dyslexia has focused on the magnocellular pathway (previously termed "transient" pathway). The magnocellular and parvocellular pathways (hereafter termed M and P pathways) are anatomically defined pathways from retina to primary visual cortex V1. Feed forward information from V1 to extrastriate cortex is at least partially segregated along ventral and dorsal pathways. Signals from the M stream are routed along both these pathways, including projection to the motion-sensitive area MT in the dorsal pathway. For a review, see Wandell (1995, chaps. 5 and 6). A general expectation would be that pattern recognition in reading (letters, words) would be primarily handled by the ventral pathway, together with other forms of visual object recognition. On the other hand, there are several lines of evidence implicating M processing deficits in dyslexia, and by extension implicating a role in reading for the M pathway.

Many studies of contrast sensitivity for sine-wave gratings have demonstrated subtle deficits in dyslexic subjects, some of which have been interpreted as evidence for a M pathway deficit. In the first such study (Lovegrove, Bowling, Badcock, & Blackwood, 1980), dyslexic children and normal controls were tested at spatial frequencies from 2 to 16 cycles/degree and stimulus exposure times from 40 ms to 1000 ms. Overall contrast sensitivity was lower for the dyslexic group than for controls, with the maximum difference occurring at 4 cycles/degree and exposure times exceeding 150 ms. Ironically, these are conditions which could well have been interpreted as implicating a P pathway deficit.[i]

Nevertheless, many subsequent studies of contrast sensitivity in dyslexic subjects have been interpreted to provide support for the M-pathway deficit hypothesis. For a review of the conflicting results, see Skottun (2000b).

The research on contrast-sensitivity deficits in dyslexics does not directly assess contrast coding in reading. In an effort to bridge this gap, we measured reading speed as a function of text contrast in dyslexic subjects and controls (O'Brien, Mansfield, & Legge, 2000). (See Section 3.3 for a review of the effects of text contrast on reading speed.) We looked for two markers of deficient contrast coding in dyslexic subjects. First, if contrast sensitivity is depressed in the channels coding information used in reading, the speed-vs.-contrast curves should be shifted horizontally to the right on the contrast axis, with a corresponding increase in critical contrast values. Second, we evaluated the possibility that dyslexic subjects

---

[i]The anatomically defined magnocellular and parvocellular pathways are thought to determine psychophysical performance ascribed to "transient" channels and "sustained" channels respectively. Several studies in the 1970's established that the transient channels mediate contrast detection at low spatial frequencies for stimuli with abrupt temporal onsets or offsets, while the sustained channels mediate contrast detection at higher spatial frequencies above about 1.5 cycles/degree (Legge, 1978). The grating spatial frequencies and exposure times in the Lovegrove et al. (1980) study comprised a partial replication of the study by Legge. Based on Legge's findings, we would expect these stimuli to be detected by sustained channels presumably associated with the parvocellular pathway.

might actually read faster with a reduction from maximum contrast, as suggested by findings in a visual search study (Williams, May, Solman, & Zhou, 1995). As we expected, maximum reading speeds were depressed in the dyslexic subjects compared to controls, but there was no evidence for the two markers of abnormal contrast coding. We interpreted our results as indicating that visual contrast coding is normal in dyslexics, at least for the channels involved in reading.

Besides the work on contrast sensitivity, there have been many studies of associations between dyslexia (and poor reading more generally) and psychophysical performance on tasks said to be mediated by the M pathway. For example, Demb, Boynton, Best, and Heeger (1998) found a correlation between reading speed and motion-discrimination thresholds (at low luminance) in a small group of adult dyslexic readers. Cornelissen et al. (1998a, b) have compared good and poor readers (both children and adults) and shown that the good readers are more sensitive to detection of motion coherence, a task thought to reflect processing in the M pathway or its projection to area MT (medial temporal area). A postmortem study of the anatomy of the lateral geniculate nuclei of five dyslexic individuals indicated that the neurons in the magnocellular layers were abnormally small, consistent with slower temporal processing of visual signals (Livingstone, Rosen, Drislane, & Galaburda, 1991). There are also brain-imaging data (fMRI) demonstrating reduced motion responses in area V5/MT of dyslexic subjects (Eden et al, 1996), and a correlation between reading speed and reduced activation of MT and adjacent motion areas in dyslexic subjects (Demb, Boynton, & Heeger, 1998).

Given the inherent variability of the dyslexic population, and the lack of agreement on diagnostic criteria or subtypes, it is not surprising that the psychophysical findings are mixed and often inconclusive regarding the hypothesized link between dyslexia and M-pathway deficits. One result has been a lively debate in the literature over the interpretation of the empirical facts. For detailed reviews see Stein and Walsh (1997), Stein, Talcott, and Walsh (2000) and Chase, Ashourzadeh, Kelly, Monfette, and Kinsey (2003) who favor the M deficit hypothesis, and Skottun (2000a, b) who opposes it.

The M pathway is thought to be specialized for motion processing or rapid temporal processing, sensitivity to low-contrast stimuli, and coarse shape (low spatial frequencies.) What could be its role in reading? Breitmeyer (1980) made an interesting and provocative proposal that spurred subsequent research. He argued that visual information about letters and words picked up in an eye fixation in reading is encoded by the P pathway (originally called "sustained" channel). But the "sustained" activity in this pathway must be erased prior to the next fixation to avoid masking of new information on the next fixation. The saccade acts as a trigger to the M system ("transient" mechanism) to inhibit or turn off the sustained activity in the P pathway. If the M system has a deficit, it would be ineffective in suppressing the sustained signals during saccades, and the result would be between-fixation masking. There are at least two problems with this model. First, it is inconsistent with high-speed RSVP reading in which there are no saccades to trigger suppression of the P-pathway signals. Second, Burr, Morrone, and Ross (1994) determined empirically that saccadic suppression is selective to signals within the M pathway (not the P pathway) and appears to occur prior to the stage of masking.

The demise of the Breitmeyer model has opened the door for other ideas about the role of the magnocellular pathway in reading. Cornelissen, Hansen, Hutton, Evangelinou, and Stern (1998), and Cornelissen, Hansen, Gilchrist, Essex, and Frankish (1998) have proposed that signals coding the relative position of letters in words is carried by the M pathway. If so, one would expect that people with deficient M processing, as indexed by high thresholds in a task involving detection of motion coherence, should experience more letter-position errors. Cornelissen et al. designed two tasks whose performance required accurate encoding of letter position. In the lexical decision task, subjects were given real words or anagrams created by interchanging pairs of letters in real words. People who have less accurate letter coding should be more likely to interpret an anagram as its corresponding word, (e.g., interpret "stnad" as "stand"). Sure enough, they found that subjects with higher motion-coherence thresholds tended to make more lexical-decision errors. While these results are encouraging, it remains to be understood why letter position would be encoded by one pathway (M pathway) and letter identity by another pathway (P pathway) and how these two critical pieces of information are ultimately bound together prior to lexical matching.

It has even been hypothesized that the M pathway is the primary pathway for reading (Chase et al., 2003). These authors tested this hypothesis using the observation from physiology that magnocellular function is suppressed by red light, predicting that normal reading should also be suppressed by red light. In R4 (1986), we reported data that test this prediction. As described in more detail in Section 4.7, we measured reading speed as a function of text color (colored letters on dark backgrounds and dark letters on colored backgrounds). The luminance-matched colors (red, green, blue and gray) were produced by combinations of neutral-density and Wratten filters overlayed on a display monitor. We found that for normally sighted subjects, reading at photopic luminances and character sizes well above the acuity limit, there was no systematic effect of color condition on reading speed; in particular, reading performance was not depressed in the red.

These findings appear to disconfirm Chase et al.'s prediction of depressed reading performance for red-illuminated text. They too found no significant effect of color on reading speed or reading comprehension of their normally sighted subjects. They did, however, find a small but significantly greater number of word errors in their red-text condition compared with other conditions. They interpret this finding to support the hypothesis of a M-pathway role in reading.

The human ability to read equiluminant text also argues against a primary role for the M pathway in reading. Cells in the M pathway are comparatively insensitive to pure chromatic contrast (e.g., a red/green border with equal luminance on both sides of the border). If strong signals in the M pathway are necessary for reading, one would expect that equiluminant text would be very hard to read. Our results (R11, 1991) and those of Knoblauch, Arditi, and Szlyk (1991) show clearly that equiluminant text with high chromatic contrast can be read at speeds matching those for text with high luminance contrast. It is likely that the equiluminant text generates some residual activity in the M pathway, due to weak luminance artifacts in the stimulus or to variations in the spectral characteristics of M-pathway cells. Yet, it would be surprising if these weak residual signals would be sufficient for normal, fast reading if the M pathway is the primary visual pathway in reading. Stein and colleagues (see Stein & Walsh, 1997; Stein et al., 2000) have argued that M-pathway deficits could manifest themselves in a variety of ways across the population of dyslexics—problems with binocular fixation stability, poor motion perception or contrast sensitivity, etc.

The idea that a M-pathway deficit might produce a wide range of subtle perceptual effects, including impact on reading, resembles another general view of dyslexia. Instead of ascribing dyslexia to a phonological deficit, a visual deficit, or both, perhaps it is a consequence of a more general temporal processing deficit. Paula Tallal has advanced the hypothesis that developmental dyslexia and some other forms of language-learning impairment are due to neural deficits in processing rapid temporal sequences such as speech signals and reading signals (Tallal, 1984; Tallal & Curtiss, 1990). Presumably, this deficit would affect the M pathway in vision more than the P pathway because of the M pathway's higher temporal bandwidth, greater sensitivity to motion, and generally faster neural conduction velocity. This hypothesis does not postulate a direct role for magnocellular signals in reading. It leaves open the possibility that magnocellular deficits in dyslexia play no causal role in reading disability, but are correlated with causal deficits in phonological processing. Farmer and Klein (1995) have provided an in-depth review and critique of evidence for the temporal-processing deficit hypothesis. They point out that there is not yet clear evidence for correlated deficits in temporal processing across sensory modalities (vision, audition, touch) as predicted by the hypothesis. Two additional limitations are that the hypothesis does not specify a physiological basis for the deficit, nor is it specific about the impact of the temporal-processing deficit on reading.

Despite the consensus that most dyslexics manifest some form of phonological deficit, it seems clear that many also reveal subtle visual deficits, typically in psychophysical tasks that rely on fast temporal processing or precise timing. It remains controversial whether these deficits are primarily confined to the M pathway from retina to cortex and to extrastriate areas specialized for motion processing. It also remains to be resolved whether the M pathway plays a specialized role in reading. It remains possible that deficits in magnocellular visual processing are correlated with but not causally connected to a core phonological processing deficit in dyslexia.

Finally, it is also possible that a specific M-pathway deficit or general temporal processing deficit could impact reading, not in its mature form, but during learning to read. The deficiency might prevent construction of the linkage between visual input and an internal lexicon or other linguistic representation.

Effect of Age on Normal Reading Speed

In R10 (1991), we compared the reading speeds (drifting-text method) of a group of 16 young participants (mean age 21.6 years) with a group of nine old participants with normal vision (mean age 67.8 years). For character sizes of 0.3°, 0.5°, and 1°, all within the optimal range for reading. Both young and old groups exceeded 300 wpm and showed no significant differences. Small deficits in reading speeds for the older participants for very tiny and very large print were attributed to age-related deficits in contrast sensitivity for very low and very high spatial frequencies. The major message from this study was that old participants who were free of eye disease read as fast as young participants in the range 0.3° to 1°.

In apparent disagreement, other studies have found slower reading in older normally sighted participants (E. M. Fine, Kirschen, & Peli, 1996; E. M. Fine et al., 1997). A recent study in our lab also found an effect of aging on reading speed. Sass, Legge and Lee (2006) used the flashcard method to measure reading speeds in groups of young and old participants, subdivided into groups with normal and low vision. We found that the old normal participants read more slowly than the young normal participants, averaging 67% of the young normal reading speed. In another study, Hartley, Stojack, Mushaney, Kiku-Annon, and Lee (1994) found mixed results. Their younger and older groups did not differ significantly in self-paced reading speeds for short essays and single sentences. But young participants were faster than older participants in a measure of the minimum exposure time for successful encoding of the underlying meaningful propositions in a sentence.

How can we explain the young/old differences in these studies given the lack of a difference in R10 (1991)? The discrepancy may be resolved by considering screening criteria and the increasing probability of sub-clinical visual or cognitive deficits in old age. In both our studies (R10, 1991; Sass et al., 2006), older participants were enrolled, based on self reports of good eye health and lab screening tests verifying letter acuity and contrast sensitivity in the normal range. But in R10, we further screened our older participants, based on detailed clinical reports. We were able to verify that only nine of the entire group of 19 older participants had completely healthy vision, that is, had no ophthalmologically apparent signs of early cataract or other symptoms of eye disease. We termed this group the "old normal group." It is this highly screened group that did not show a statistically significant difference from the young normal group. However, in R10, we also reported that a larger subset of our entire sample of older participants, including several participants with sub-clinical eye disorders (cataract and retinal detachment), did read significantly slower than the young normal group. In the Sass et al study, the acceptance criteria for older participants included self reports of good eye health (no known eye disease) and laboratory measurement of letter acuity and contrast sensitivity in the range for normal vision, but not the stringent clinical screening used to define the old normal group in R10. It is likely that the group of older participants in the Sass et al. study was comparable to the less stringently screened larger group in R10, and in both cases, these groups read significantly slower than the corresponding young normal groups.

The preceding paragraph makes the point that age differences in reading speed may be attributable to subtle visual or cognitive deficits, even in participants with no diagnosed eye disease. Strong corroboration for this point comes from a large population-based study. Lott et al. (2001) obtained reading speeds and many other measures of visual and nonvisual function from 900 participants. Reading-speed data were analyzed from a subsample of 544 participants with good acuity (better than 20/32). Their ages ranged from 58 to 102 years. Reading speed was found to decline with age, but multiple-regression analysis indicated that when other measures were taken into account (especially low-contrast acuity, a measure of motor ability, and a measure of attention), age was no longer an independent predictor of reading speed. An interpretation of these findings is that reading speed declines in old age, even for people with normal visual acuity and self-reported healthy vision. But it is likely that much of the measurable decline can be traced to subtle visual, motor or cognitive deficits, with the prevalence of these subtle deficits increasing with age. It remains for future research to determine the precise nature of these subtle deficits and how they influence reading speed.

The general point here is that subtle visual and cognitive factors appear to play a part in explaining individual differences in reading speed. If the prevalence of such deficits is higher in one group, such as old participants, than another group, such as young participants, we are likely to find group differences. If the relevant factors are equated between the groups, the differences may diminish or disappear.

Normally sighted children reach nearly adult levels of visual acuity by age 7 (Dowdeswell, Slater, Broomhall, & Tripp, 1995). By first grade, most of them know the alphabet. Nevertheless, reading speed takes a long time to reach adult levels. Carver (1990) reports data indicating that mean reading speed increases about 14 wpm for each year in school from first grade to the fourth year of college. For example, mean reading speeds in 4th, 8th, and 12th grades are reported to be 158, 204 and 250 wpm respectively. It remains unclear what mix of developmental factors in perception and cognition account for this gradual growth of reading speed.

Low-Vision Reading Speed

When text properties are degraded, or eye disease disrupts vision, there is no question that visual factors can limit reading speed. Almost everyone with low vision has difficulty reading. One definition of low vision is the inability to read the newspaper with best refractive correction (glasses, or contacts) at a normal distance of 40 cm (16 in.). Leat, Legge, and Bullimore (1999) reviewed studies showing that the primary presenting complaint at low-vision clinics is typically problems with reading.

From a practical point of view, there are two components to reading difficulty in low vision: the reduced range of print sizes that are legible, and the speed of reading. We review the issue of print size in section 3.2. Here, we ask a very basic question: If print size is sufficiently magnified, how many people with low vision can achieve normal reading speed?

Although a definitive answer to this question would require a large population-based study of low-vision reading speed, a preliminary answer is available from our study of 141 patients entering the low-vision clinic of the Minneapolis Society for the Blind (R12, 1992). In this study, we used the flashcard method to measure reading speeds for highly magnified text. The letters each subtended 6°, about 70 times larger than 20/20 letters, and well above the critical print size (CPS; see section 3.2 for a definition) for most people with low vision. We found that only 30% of the low-vision participants could achieve a threshold criterion of normal reading speed, defined as two standard deviations below the mean of a normally sighted control group (R12, 1992, Fig. 4). These findings show that magnification per se is often insufficient to overcome low-vision reading problems. We return to the factors limiting low-vision reading performance in chapter 3.

When people with low vision achieve maximum speeds that are below normal, we can view the glass as half full or half empty. The reduced reading speed can certainly pose problems for education, work, or recreation. But it is important to realize that people with severe reading problems can benefit from visual reading, even if high magnification is required, and even if reading is slow.

Whittaker and Lovie-Kitchin (1993) have identified three subnormal levels of reading performance. These include (a) "spot reading" (44 wpm) adequate for many activities requiring brief reading such as dealing with mail, recipes, and medicine instructions; (b) "fluent reading" (88 wpm); and (c) "high fluent reading" (176 wpm). Although the "high-fluent rate" is well below the typical normal value (250–300 wpm), sustainable reading at this rate is highly functional.

Some people with low vision may only achieve the spot or fluent levels, and may be able to sustain these rates only for short periods of time. Visual reading is still effective for some tasks, but lengthy texts may require nonvisual methods of reading.

Auditory and Tactile Reading Speed

Many people who are blind or have low vision read by sound or touch. Screen readers allow digital document files on computers to be read aloud by synthetic speech. Two popular screen-reading programs are Jaws (Freedom Scientific, St. Petersburg, FL: http://www.freedomscientific.com) and Window-Eyes (GW Micro Inc., Fort Wayne, IN: http://www.gwmicro.com).

How quickly can people read with synthetic speech? Hensil and Whittaker (2000) asked participants to read calibrated passages visually as quickly as possible for comprehension (silent reading). After each passage, they answered comprehension questions. The same participants listened to passages presented by synthetic speech (DECTalk synthesizer[10]), at increasing rates, followed by comprehension questions. Maximum auditory rate was determined by the speech rate at which comprehension dropped to a criterion of 2 out of 3 questions correct. The mean visual reading rate (30 sighted participants) was 246 wpm, and the mean auditory rate was 280 wpm.[11] Carver (1982) measured comprehension performance for text presented at controlled rates both for vision (motion picture film) and for hearing (time-compressed speech recordings). From his analysis, he identified the optimal rates for vision and hearing, defined by the maximum amount of information comprehended per unit time. He found that the optimal rate of presentation was the same for vision and hearing, about 300 wpm. The bottom line from these two studies is that auditory limitations on reading speed appear to be very similar to visual limitations.

Braille is the most successful code for reading by touch. Box 2.2 briefly summarizes some of its important characteristics.

The conventional estimate of Braille reading speed is about 100 wpm (Foulke, 1982). This value is controversial because studies using different methods and materials have arrived at widely disparate estimates. In an effort to standardize measurements, we developed a variant of the MNREAD test to measure Braille reading speed (Legge, Madison, & Mansfield, 1999).

Figure 2.3 shows reading speeds for 44 Braille readers. The graph also replots print reading speeds from R15 (1996) for 50 normally sighted participants, and 39 low-vision participants. All speeds were measured with nearly identical MNREAD sentences and procedures. The graph shows a reading speed for each participant, plotted against their percentile position within their group: lowest on the left and highest on the right.

The filled symbols show the reading speeds for the Braille readers. The median speed (50th percentile) is about 122 wpm, with a range from 24 to 234 wpm. These values correspond to a median rate of 7.5 characters per second on the fingertip, with a range of 1.5 to 14.4 characters per second. We observed that uncontracted

---

[10]The DECTalk synthesizer has high intelligibility. In the same study, one of the authors recorded passages at different speaking rates for comparison with DECTalk. The results with the passages read by human voice were very similar to those with DECTalk.

[11]The slightly slower speeds for visual reading might have been due to a criterion effect; the visual reading task, which relied on self pacing, may not have been as demanding as the auditory task in which presentation rate increased until comprehension deteriorated (S. G. Whittaker, personal communication, January 15, 2003).

Box 2.2 Characteristics of Braille

Figure 2.2 provides an overview of the Braille code and includes examples of sentences in contracted and uncontracted Braille.

Braille is not a language like English or Spanish, but a code for representing the written form of a language with tactile symbols. (Similarly, Morse code is a method for representing the written form of a language with dots and dashes.) Braille characters consist of combinations of six dots, placed in a matrix of two columns and three rows, termed a Braille "cell." The 64 possible arrangements of dots in the Braille cell (including the blank cell) comprise the 64 characters of the Braille code. Twenty-six of these characters represent the 26 letters of the English alphabet (Fig. 2.2). Rather than designating different Braille characters for uppercase and lowercase letters, a capitalization marker preceding a letter, indicates that it is uppercase. Several additional Braille characters are used for punctuation marks. Numbers are represented in Braille by using the first 10 letters of the alphabet to represent the digits 1, 2, ... 9, 0. Numbers are distinguished from letters by a number sign character.

Braille has a contracted version and an uncontracted version (Fig. 2.2). In uncontracted Braille, sometimes termed Grade 1 Braille, there is nearly a one-to-one correspondence between printed characters and Braille characters. (An exception is the capitalization character which does not appear in print.) Contracted Braille, sometimes termed Grade 2 Braille, is the standard version used in most books, magazines and other literary material. Contracted Braille contains about 192 contractions. Figure 2.2 gives examples of several types of contractions. Notice that some of the 64 Braille characters, not used as letters or punctuation marks, are used as contractions, e.g., separate characters for "the," "for," and "ing." The contractions serve three purposes: to reduce the bulk of Braille, to increase reading speed, and to reduce writing time, a particular issue for those who write Braille dot by dot with a slate and stylus. The contractions reduce the length of Braille texts by about 25 percent (Legge, Madison, & Mansfield, 1999).

Braille was invented, almost single-handedly, by Louis Braille (1809–1852). He was blinded in an accident at the age of 3, and later studied and taught at the National Institution for the Blind in Paris where he invented and wrote about the Braille code (Mellor, 2006). One mark of his genius in designing the system is that the basic elements of the Braille code (the Braille cell, its spatial properties, and the Braille alphabet) are fundamentally the same as the system introduced by Braille. Acceptance of the Braille code has had an interesting and sometimes controversial history. For instance there was debate between advocates of dot codes (such as Braille) in which the characters bear no shape resemblance to printed characters, and advocates of embossed characters that are similar in shape to printed characters. For historical reviews, see Farrell (1956) and National Library Service (2000). For an experimental comparison of the legibility of Braille letters and printed letters, see Loomis (1981).

Alphabet.	Standard dimensions of Braille.

Alphabet.

a b c d e f g h i j

k l m n o p q r s t

u v w x y z

Standard dimensions of Braille.

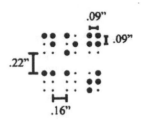

.09"

.09"

.22"

.16"

Capitalization. A capital letter is preceded by a capitalization marker (a dot in the lower right corner of the Braille cell) as in *Mary*

Numbers. Digits *1, 2, 3, 4, 5, 6, 7, 8, 9, 0* are the letters *a-j*. Numbers are preceded by a number sign (e.g. 123)

Punctuation. There are Braille characters for punctuation marks as in *dog,* and *cat.*

Contractions. There are 192 contractions, such as:

Word Contractions

knowledge	people	the	and	for	good	father	under

Contractions for Common Letter Combinations

ed	ing	th	ar	st	er

A sentence shown in print, uncontracted Braille, and contracted Braille.

Printed Version
The two friends did not know what time the play would start.

Uncontracted Braille Version

Contracted Braille Version

Figure 2.2   In the Braille examples in this figure, the six possible dot positions in each Braille cell are portrayed in outline with the actual dots shown in black. For instance, the Braille character consisting of the top two dots of the cell represents the letter *c*. The top left panel shows the Braille alphabet, and the top right panel shows the standard dimensions of the Braille cell. According to Nolan and Kederis (1969), individual dots have a base diameter of 0.060 in. (1.5 mm), and a height above the page of 0.017 in. (0.43 mm). The middle panel shows several features of the Braille code, including examples of the 192 contractions. The lower panel shows a printed sentence and its corresponding uncontracted and contracted Braille versions.

Braille was read at 71.5% of the speed of contracted Braille, when measured in wpm, but this difference was almost entirely accounted for by the difference in the number of characters. Similar to Carver's (1990) invariance principle for print reading speed, Braille reading speed for the contracted and uncontracted codes is constant in characters per sec.

Open circles in Figure 2.3 show data for the normally sighted participants. Their median is about 250 wpm, compared to about 122 wpm for the Braille readers.

Why is Braille reading usually slower than print reading? Foulke (1982) answered this question by pointing out that only about one character is recognized at a time on the fingertip. If so, we might expect Braille reading speed to be equivalent to print reading speed when only one print character is visible at a time. In R1 (1985), we used the drifting text method to measure reading speed as a function of "window size," the number of letters visible on the screen at one time. (See section 4.4 for a review of window-size effects.). The drop-off in print reading speed when we reduced the window to just one character was about a factor of two, just like the difference in medians between print and Braille reading speeds in Figure 2.3. This finding supports Foulke's view that Braille reading speed is limited by the small tactile span of about one character.

The open triangles show reading speeds for a group of low-vision participants, most of whom had macular degeneration. The distribution of low-vision speeds was similar to our Braille readers; median of 120 wpm with a range from 12 to 264 wpm. This similarity may suggest that Braille is a good alternative for people with low vision. But keep in mind that low vision often occurs late in life when it is difficult to learn Braille.

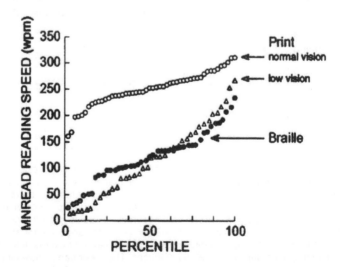

Figure 2.3 Reading speeds (in wpm) are shown for three groups: 44 Braille readers (filled circles) studied by Legge, Madison, and Mansfield (1999), and groups of 50 normally sighted (open circles) and 39 low-vision (open triangles) print readers studied in R15 (1996). See the text for more details (from Legge et al., 1999, Fig. 3.)

Is Reading Speed Limited by Motor Movements, Sensory Coding, or Cognition?

For difficult text with challenging vocabulary or complex ideas, it is not surprising that reading slows down (cf. Rayner & Pollatsek, 1989, chap. 4). Under such conditions, reading speed is limited by comprehension demands. But for linguistically simple texts, perceptual and motor factors are more influential in limiting reading speed.

Carver's (1976, 1990) discovery that reading speed is independent of text difficulty (provided the difficulty does not exceed the reader's grade level) has an important theoretical consequence. It implies that reading speed is not limited by lexical access or other linguistic processes. This assertion appears to conflict with the well-known results of Just and Carpenter (1980). These authors described a model of how a meaningful representation is built in working memory from online analysis of text in reading. Just and Carpenter (1980) showed that a variety of linguistically-based predictor variables (e.g., word frequency, number of syllables, and grammatical part of speech) influenced the time to read individual words and phrases in text. The model did not include any visual constraints on the reading process. Carver (1990, Appendix A) has presented a reanalysis of some of the Just and Carpenter (1980) results, and finds that number of standard-length words by itself accounts for most of the variance in reading times ($R = 0.98$). This re-analysis does not deny the importance of linguistic attributes of text in building cognitive representations, but does imply that either reading speed is limited by perceptual or oculomotor factors, or that the many variables identified by Just and Carpenter (1980), and studied by other researchers, combine to produce reading speeds that are invariant with text difficulty.[12]

Carver's (1976, 1990) invariance principle has an important implication for the use of reading speed as a psychophysical measure; it means that for linguistically simple texts, experimenters need not worry about contamination of performance by uncontrolled variability in the linguistic difficulty or content of test material. The only linguistic impact on reading speed can be taken into account by using a character-based metric for measuring reading speed, such as characters per sec or slwpm.

Carver's (1976, 1990) invariance principle is not universal. As discussed in section 2.2, there is a robust "context effect" in which continuous text is read faster

---

[12]Keep in mind that reading speed is a global measure, computed over a block of text. The gaze-duration measures of Just and Carpenter (1980), and many other studies of variation in eye-fixation times, are local oculomotor measures. It is possible that trade-offs at the local level of analysis can reconcile Carver's (1976, 1990) rate invariance principle with local linguistic influences on fixation times. For example, imagine two blocks of text matched for the number of characters, but one more difficult than the other. The more difficult text will typically have longer mean word length, and fewer actual words. The longer words will have lower word frequencies, since word frequency decreases on average with increasing word length. People fixate on most words in reading, but lower frequency words require longer fixation times (Rayner, Sereno & Raney, 1996). In our example, the more difficult text will have fewer words to process, but the words will have longer fixation times on average. The net effect may be approximately equal reading time (equal reading speed) for the two passages. (This example of a plausible tradeoff at the local level was suggested to me by Bosco Tjan.)

than scrambled text or random strings of words, matched for the number of characters. If reading speed were entirely independent of interword linguistic constraints, and strictly dependent on the number of characters, no such effect should occur. Perhaps the gross difference between continuous text and random strings is large enough to have a measurable effect on reading speed, whereas more subtle differences due to variations in text difficulty (or mean word length) are too small to empirically detect.

Normal visual page reading is limited to about 300 wpm by a ceiling imposed by eye movements. When this ceiling is removed, using RSVP, people can read at least two or three times faster. In R20, we presented a model that argues that RSVP reading speed in central vision is limited by spatial and temporal limitations of visual letter recognition. In other words, when eye-movements no longer limit reading speed, the bottleneck on reading speed is perceptual.

Under difficult stimulus conditions (low contrast, tiny letters, etc.), even normal page reading slows down due to visual factors. Many of these factors are discussed in our series of articles.

For most people with low vision, eye disease has the effect of reducing the quality of text input, resulting in reduction in reading speed below limitations associated with eye movements. Loss of spatial resolution (acuity) often necessitates the use of a magnifier. Manipulation of a magnifier can itself impose additional burdens on motor control that reduce reading speed (see section 4.6).

Braille reading speed is also perceptually limited to about 7.5 characters per second related to the one-character-wide tactile span of the fingertip.

The major point is that the input rate for processing words in text is often limited by the characteristics of the sensory channel—vision, touch, audition—or by the page-navigation mechanism including eye movements or magnifier movements.

2.4 COMPARISON OF READING SPEED TO OTHER MEASURES OF READING PERFORMANCE

Reading performance has been evaluated in many ways. We briefly comment on the relationship of reading speed to several other methods for assessing the impact of visual factors.

Comprehension

Understanding of meaning is usually the primary goal of reading. Comprehension tests, based on multiple-choice questions, are widely used in the educational assessment of reading ability and verbal aptitude. The first major empirical study of reading speed (Abell, 1894) was concerned with the link between accuracy of comprehension and reading speed; do people who read faster have better or worse comprehension? It turns out that there is no simple relationship between natural reading speed and level of comprehension. See the brief review of findings in R7 (1989).

Why not use comprehension as a psychophysical measure of reading performance? The answer is that comprehension is less closely coupled to visual processing than reading speed, and too difficult to measure accurately for parametric study of stimulus variables.

In R7 (1989), we used the drifting-text method to assess comprehension for passages presented at rates ranging from 10 wpm to 450 wpm. The passages were about 150 words in length, so the corresponding presentation times ranged from 15 min to 20 sec. There were 6 multiple-choice questions after each passage. We tested 109 normally sighted participants and 24 low-vision participants. We had two goals. The first was to assess the properties of comprehension as a psychophysical measure, and the second was to determine if very slow reading speeds, which sometimes result from low vision, were compatible with good comprehension.

We discovered three aspects of comprehension testing that makes it unsuitable for a psychophysical measure of reading performance: (a) The time taken to administer multi-choice comprehension questions and the relatively sparse amount of data resulting from them make it difficult to obtain accurate performance estimates across a range of stimulus conditions. Individual psychometric functions of comprehension accuracy versus drift rate are much noisier than the corresponding psychometric functions for word recognition (compare Fig. 2.1[a] with Fig. 2 in R7, 1989). (b) Group data can be used to reduce the noise in individual psychometric functions. When this is done, the transition between good comprehension and poor comprehension as drift rate increases is seen to be quite gradual, rather than the sharp transition observed for percent accuracy of words read (R7, 1989, Fig. 3). Carver (1982) also documented a slow decline in comprehension accuracy with increasing presentation rate. (c) There is substantial individual variation in comprehension performance among visually normal participants, presumably due to nonvisual cognitive factors. For these reasons, in most of our studies, we did not use comprehension as a primary measure. In some studies, however, we have included comprehension questions in order to encourage participants to understand text during reading (cf. R17, 1998).

There are two reasons why comprehension might be poorer for low-vision readers. First, there has been speculation that comprehension breaks down at slow reading speeds because it is difficult to integrate meaning across words or phrases, or because it is difficult to maintain attention on the text (Gibson & Levin, 1975). Contrary to this expectation, in our study, the comprehension performance of normally sighted participants remained high, even at the excruciatingly slow presentation rate of 10 wpm. Second, the added demands of decoding poorer quality visual input resulting from eye disease might detract from comprehension. Because we found that comprehension scores for normally sighted participants began to decline as they approached their maximum reading speeds (the decline beginning at about two-thirds of their maximum rates), we tested comprehension of our 24 low-vision participants with drifting text at about two thirds of their individually determined maximum reading speeds. (They were also tested at 84% of maximum.) Contrary to the pessimistic expectations, comprehension was within the normal range for most of the low-vision participants, many of whom had maximum speeds well below 100 wpm.

Although comprehension performance may not be a good psychophysical measure of visual factors in reading, educators or rehabilitation specialists may need to measure reading comprehension in people with low vision. Commonly used tests may be too time consuming and demanding for slow readers with low vision. Watson, Wright, Long, and De l'Aune (1996) have developed a short test of reading comprehension designed for people with low vision, particularly older people with macular degeneration. The Low Vision Reading Comprehension Assessment requires only about 9 min to administer, is reliable (test–retest correlation of .871) and correlates highly with performance on the Woodcock Reading Mastery test ($R = .827$).

Teachers often need to decide whether to allow low-vision students extra time on exams. Since most people with low vision read more slowly than sighted peers, even with magnification, it is entirely appropriate to allow low-vision students to have extra time to complete exams. Given the extra time, however, it is reasonable to expect that low-vision students' comprehension of the textual material should not be compromised by their eye condition. Keep in mind, however, that tests often involve more than the reading of continuous text. Analysis of graphics, or the ability to skim for key words or phrases, or the ability to look back and forth to make comparisons, might be compromised by low vision in ways that cannot be compensated for by simply allowing extra time. We return to a related issue in section 4.6 when we discuss hypertext search and retrieval.

Regarding testing conditions for visually impaired students, it should also be noted that reading ability may be much more sensitive to test-taking conditions. As the papers in our series make abundantly clear, low-vision reading can be abnormally sensitive to attributes of the text, such as contrast, page layout, character size, and also to viewing conditions such as light level, or glare.

For people with low vision who read very slowly (< 100 wpm), it is impractical to read novels or other lengthy texts. For prolonged reading, it may be more efficient to rely on auditory or tactile reading. But reading short passages visually can be very important. People with low vision need to read letters and memos, recipes, medicine bottles, check books and personal records, telephone numbers and short newspaper articles. From R7 (1989), we now know that normal comprehension is possible for such tasks, even if reading is very slow. This finding emphasizes the importance of prescription of appropriate low-vision reading aids, even in cases of severe visual impairment.

Eye Movements

There is a vast literature on reading eye movements in static text, mostly based on normally sighted participants. Key contributors to this literature during the past 25 years and a few of their key writings include George McConkie (McConkie, Kerr, Reddix,

& Zola, 1988; McConkie, Kerr, Reddix, Zola, & Jacobs 1989; McConkie & Rayner, 1975), Keith Rayner (1998; Rayner & McConkie, 1976; Rayner & Pollatsek, 1989;), and Kevin O'Regan (1990; O'Regan, Levy-Schoen, & Jacobs, 1983).

The pattern of reading eye movements in static text consists of a sequence of saccades separated by fixations. Key issues in the literature on eye movements in reading include the amount of visual information that is encoded on each fixation, the nature and extent of online transformation of visual input into linguistic representation, factors governing the planning and timing of saccades, and the mechanics of oculomotor control.

Reading speed is approximately equal to the mean saccade length divided by the average fixation time.[13] For example, a person who reads 300 slwpm is reading 1,800 characters per minute or 30 characters per second. If they are making 4 to 5 eye fixations per second, they are advancing on average 6 to 7 new characters per eye movement. During one hour of reading they recognize about 108,000 characters.

When reading speed slows down due to deficient visual input (either degraded text quality or visual impairment), the reduction could be due to prolonged fixations, shorter saccades, or some combination of the two. There is a rough consensus in the literature on normal reading that cognitive factors have a greater impact on the variability of fixation times, while visual and oculomotor factors exert more impact on saccade length (cf. Starr & Rayner, 2001). Consistent with this consensus, several studies of eye movements in people with central scotomas from macular disease have found that slower reading is primarily due to abnormally short saccades, while fixation times are more nearly normal (Bullimore & Bailey, 1995; Rumney & Leat, 1994; Trauzettel-Klosinski, Teschner, Tornow, & Zrenner, 1994).

Why should deficient visual input result in shorter saccades? In section 3.7, we argue that both impoverished text and visual impairment result in reduction in the size of the visual span, the number of letters that can be recognized reliably without moving the eyes (see also R16, 1997; R20, 2001). When the visual span is reduced in size, fewer letters can be recognized on each fixation. Readers must make shorter saccades to move through text in order not to miss words or letters between fixations. The characteristics of the visual span appear to impose a fundamental sensory bottleneck on reading performance.

In short, when reading speed slows down because text is hard to see or because of visual impairment, the primary impact on eye movements is likely to be a reduction in the length of saccades. If reading slows down because the meaning of the text is difficult to understand, we may also observe an increase in fixation times.

Reading Accuracy, Visual Search, and Visual Comfort

Accuracy refers to the proportion of words read correctly. In the drifting-text and flashcard methods (section 2.1), the calculation of reading speed includes a cor-

---

[13]This approximation does not take into account the finite time for saccades, and the extra saccades associated with finding new lines or new columns of text.

rection for errors, that is, accuracy is incorporated into the measure. In the RSVP method, the level of accuracy is used as a criterion for establishing a threshold exposure time for reading.

It is sometimes desirable to separately measure reading time and accuracy. For example, when participants read static text with unlimited viewing time, such as printed text on paper, they adopt a criterion that balances speed against accuracy. Although instructions can be used to encourage participants to use a consistent criterion, it is possible that reading speed, accuracy or both change as a function of some stimulus parameter. For instance, in a recent study of the impact of color on reading performance, Chase, Ashourzadeh, Kelly, Monfette, and Kinsey (2003) found that red light impairs reading accuracy, but not reading speed. (We discuss color effects on reading in section 4.7.)

Reading accuracy is occasionally considered as a separate variable in testing low vision as well. For instance, the Pepper Visual Skills for Reading test (Baldasare, Watson, Whittaker, & Miller-Shaffer, 1986) includes an accuracy score, which is intended to evaluate the ability of low-vision participants to proceed serially through text without missing words.

Visual Search has been used frequently to measure visual factors in reading. For a recent example, see Scharff, Hill, and Ahumada (2000). They studied the effect of text contrast and characteristics of the background. (The text was overlaid on a textured background.) Participants searched through the text for a target word.

Visual search has rarely been used to study low-vision reading. One exception is our study of hypertext retrieval (R19, 2002). Because hypertext retrieval involves a skimming strategy that differs from typical line-by-line text reading, visual search is a natural task for assessing hypertext reading. In our study, normal and low-vision participants were asked to search through text-only websites for the answers to specific questions. One purpose was to assess the impact of the spatial arrangement of hyperlinks. We measured the time taken to find the answers to the test questions, and the paths taken (i.e., sequence of pages accessed). We discuss the results of this study in section 4.6 in relation to potential problems with hypertext reading for people with low vision.

Roufs and Boschmann (1997) used an interesting variant of visual search to study image quality of computer text displays. They created displays of "pseudo-text," random strings of characters approximating the distribution of word lengths and spacing in real text. Participants searched the pseudotext for a target letter. The main purpose for using pseudotext was to eliminate all linguistic meaning, while retaining a good surrogate of a text stimulus. The pseudotext method may be promising for evaluating reading-like performance for participants when test materials in the participant's language are not available. It should be noted, however, that there is debate about the similarity of eye-movement patterns in pseudotext search and reading. Vitu, O'Regan, Inhoff, and Topolski (1995) emphasize the similarity and hence the lack of influence of top-down linguistic processes on oculomotor control in reading, but

Rayner and Fischer (1996) emphasize differences in fixation times, word skipping, and refixations.[14]

Roufs and Boschmann (1997) also described a method in which participants judged the "visual comfort" of text displays on an integer scale from 1 (*poor*) to 10 (*excellent*). Visual comfort is the psychological variable, measured by the participant's judgment of the "ease with which the information can be read from the screen." Visual comfort undoubtedly depends on well-studied text variables such as contrast and character size. It may also vary with page layout, color of foreground and background, and any other factors that may influence a participant's psychological response. Roufs and Boschmann showed that visual comfort is highly correlated with search time for texts displayed with different contrasts.

Reading conditions that result in eye strain, sometimes termed "asthenopia, will undoubtedly influence visual comfort. Sheedy, Hayes and Engle (2003) asked 20 normally sighted participants to perform eight reading tasks with viewing conditions likely to induce discomfort, e.g., close viewing distance, small font, flicker, and glare. The participants then filled out a survey aimed at determining the nature and extent of their discomfort. The authors used factor analysis to identify two general factors contributing to eye strain in normal vision—external ocular surface irritations such as dry eye, and "behind-the eye" muscular factors including convergence and accommodation demands. Both of these factors can be exaggerated in low vision. In addition, for people with low vision, the ergonomic demands of handling a magnifier may be fatiguing or cause discomfort to the back, neck, shoulder, arm, wrist or hand. Not surprisingly, many people with low vision complain of fatigue or discomfort when reading.

Visual comfort may determine reading endurance for people with low vision. Many visually impaired people are unable to sustain reading beyond a limited period, sometimes only a few minutes. A person may have good reading speed and comprehension, but very limited endurance.

There is a need for research to develop accurate and convenient predictors of reading endurance for people with low vision. Consider one possible way of assessing the joint effects of speed and endurance on reading function. If visual discomfort limits the length of a reading session to T minutes, for a person with a reading speed S wpm, then the total number of words covered in the session is the product of S and T. For instance, if the reading speed is 100 wpm, and the reading time is 10 min, then the person can cover a maximum of 1,000 words in a single session.

---

[14]Unlike Roufs and Boschmann (1997) who used strings of different letters, Vitu, O'Regan, Inhoff, and Topolski (1995), and Fischer (1996) created pseudotext by replacing all the letters of real passages with z's.

3
Visual Mechanisms in Reading

Gordon E. Legge

What are the roles of spatial-frequency channels and other visual mechanisms in reading? Most previous research on reading has focused on cognitive and linguistic influences, but our research has highlighted the importance of sensory factors. Our work on the reading difficulties of people with low vision shows how reading breaks down in the presence of sensory loss, providing a revealing contrast with normal visual function.

This chapter emphasizes the impact of three aspects of visual processing on reading—contrast coding, size coding, and the decrease in spatial resolution outward from the fovea. In the first three sections of this chapter, we consider the impact of character size and contrast on reading. In sections 3.4 and 3.5, we link these psychophysical findings to the human contrast sensitivity function (CSF) and the representation of patterns by spatial-frequency channels. In section 3.6, we discuss reading in peripheral vision. We are interested in peripheral vision for two reasons: because it reveals a limitation of the CSF model of reading, and because people with central-field loss must rely on peripheral vision to read.

In section 3.7, we discuss the visual span in reading, a unifying concept that helps to bridge the gap between the basic psychophysical findings and reading speed. The visual span refers to the number of letters, arranged side-by-side as in text, that can be recognized accurately without moving the eyes. We argue that the size of the visual span is limited by sensory factors, and imposes a bottleneck on reading speed in both normal and low vision.

In section 3.9, we discuss a model that uses the concept of the visual span to forge a link between letter recognition and reading speed.

This chapter also shows how the major categories of low-vision reading deficits can be understood in terms of deficient sensory processing. The three key stimulus variables discussed in this chapter—character size, contrast, and retinal eccentricity—correspond to a common classification of reading problems in low vision in terms of low acuity, reduced contrast sensitivity, and visual field loss. Central-field loss is particularly important because it is so common in older people and often so deleterious to reading. We consider the characteristics of these three types of low-vision reading deficits in sections 3.2, 3.3 and 3.8.

In chapter 1, we optimistically predicted that researchers are on the threshold of a major accomplishment—a fully elaborated theory of visual-information processing in reading, linking image formation on the retina to cognition. We hope that the research reviewed in the present chapter contributes to this goal by characterizing the visual front end of reading.

The first stage in reading is the optical transformation from characters on a page or screen to a retinal image. This raises the question, is reading limited by the optics of the eye? Blurry reading often prompts people to visit an eye specialist. For most people, the problem is refractive error (myopia, hypermetropia, and astigmatism) which can be corrected with glasses, contact lenses or refractive surgery. What limitations on reading are imposed by optical image formation once refractive errors have been corrected? In the healthy human eye, optical image quality in central vision is adequate to support higher letter acuity than the best values measured for human participants. Beckmann and Legge (2002) developed an ideal-observer model to estimate the best possible letter acuity in central and peripheral vision, given the known characteristics of retinal image formation in the human eye. This model followed in the steps of Geisler's (1989) sequential ideal observer model and dealt with the initial pre-neural stages of letter recognition. The model took into account the pixel rendering of characters on a computer screen, retinal image formation by the optics of the normal human eye, and the pattern of photon absorption by the retinal cone mosaic. Following these stages of encoding, the model's letter-recognition decision was determined by an optimal algorithm for classifying stimuli as one of 26 lowercase letters. Beckmann and Legge compared the performance of this ideal-observer model to human performance in central vision and in peripheral vision at 5° and 20° of eccentricity. In central vision, the model's letter acuity was seven times better than human acuity for high-contrast letters, and at 20° eccentricity, the model's acuity was 50 times better. These results show that preneural factors in normal human vision are not the limiting factors for letter recognition or reading. We must search for limitations on reading performance at later stages of visual processing.

3.1 TWO IMPORTANT PROPERTIES OF LETTERS IN TEXT: CONTRAST AND SIZE

Before reviewing the dependence of reading speed on character size and contrast, we briefly review some key definitions. Psychophysical studies from the 1950s through the 1970s highlighted the importance of two dominant stimulus variables in vision—contrast and size. "Contrast" refers to luminance differences between the target and background rather than to the overall luminance levels. Both psychophysical and neurophysiological studies have demonstrated that contrast signals are extracted from retinal images for visual encoding (cf. Shapley & Enroth-Cugell, 1984). "Size" usually refers to a target's angular size, proportional to retinal-image size. Evidence for spatial-frequency channels in human vision and for retinal and cortical neurons with receptive fields varying in size make clear that target size is an important variable in early visual coding. It is natural to ask how the size and contrast of text letters affect reading performance.

Contrast Definitions

The contrast of text letters refers to the luminance difference between the strokes (or pixels) making up a letter's features and the background. "Contrast polarity" of text refers to the distinction between dark letters on a white background (normally the case for printed materials) and bright letters on a dark background ("reversed" contrast). Figure 3.1 shows lines of text in seven steps of contrast for each of the two contrast polarities.

In Figure 3.1, contrast is manipulated by changing the luminance of both the letters and the background. In typical lab experiments on the effect of contrast, the luminance of the bright background is held constant. Maximum contrast occurs for black letters on the bright background, and contrast is reduced by increasing the luminance of the letters, rendering them as shades of gray on the constant bright background. As the luminance of the letters approaches the luminance of the background, the contrast of the letters diminishes towards threshold.

There are several metrics for stimulus contrast. In most cases, we have used Michelson contrast:

$$C_{Michelson} = (L_{max} - L_{min}) / (L_{max} + L_{min})$$

where L_{max} refers to the maximum luminance (white background for black-on-white text) and L_{min} refers to the minimum luminance (dark letters). Michelson contrast ranges from 0 to 1.0, and is sometimes expressed as a percentage from 0% to 100%. Two advantages of this metric for text contrast are (a) contrast values range from 0 to 1.0 for both polarities of text; and (b) the Michelson definition is widely used for sine-wave grating stimuli, simplifying comparison with the grating literature.

Weber contrast is an alternative metric:

$$C_{Weber} = (L_{target} - L_{background}) / L_{background}$$

where "target" and "background" refer to the letters and the page (or screen). For dark letters on a bright background, Weber contrast is negative with a greatest

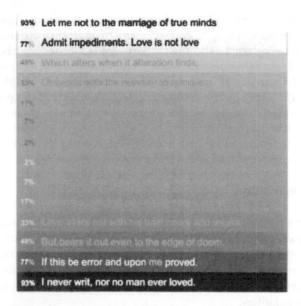

Figure 3.1 Examples of contrast magnitude and contrast polarity. The 14 lines of Shakespeare's Sonnet CXVI are rendered with seven contrast levels for each of the two contrast polarities. Numbers refer to the Michelson contrast between the letters and the background in the original photograph.

value of −1.0. For bright letters on a dark background, Weber contrast is positive, and unbounded in magnitude. Because Weber contrast is linearly related to the luminance difference between target and background, it is preferable for some types of theoretical analysis, such as the calculation of "contrast energy" used in estimating human statistical efficiency for letter recognition (Parish & Sperling, 1991; Pelli, Farell, & Moore, 2003; J. A. Solomon & Pelli, 1994; Tjan, Braje, Legge, & Kersten, 1995).

Size Definitions

We discuss three issues concerning the definition of character size: the distinction between physical and angular character size, definition of a size measure for a set of letters, and the distinction between fixed-width and proportionally spaced fonts.

First, we distinguish between physical and angular character size. Physical character size is obtained by direct measurement of the width or height of the character on the page with a ruler or equivalent. Angular character size is the corresponding visual angle subtended at the observer's eye,[1] and depends on both the physical size and the viewing distance:

---

[1]"Physical" and "angular" size are sometimes termed "object" and "retinal" size.

Angular Size in radians $\approx$ Physical Size / Viewing Distance

Angular Size in degrees $\approx$ 57.3 × Angular Size in radians
= 57.3 × Physical Size / Viewing Distance

where the physical size and viewing distance must be measured in the same units (typically, millimeters, centimeters, or inches). These approximations hold when the physical size is significantly smaller than the viewing distance. For instance, suppose the height of the lowercase letter "x" in a newspaper column is 1.5 mm (physical size), and the reader views the newspaper from a typical reading distance of 40 cm (16 inches). The angular height of "x" at the eye is about 0.2° (equals 12 min-arc). If the reader reduces the viewing distance from 40 cm to 20 cm, the angular character size doubles to 0.4°, but the physical character size, of course, does not change. Angular character size is generally used in studying vision because it determines retinal-image size. Typographers and display designers are more interested in physical size because it determines how many text characters fit on a line or screen of fixed dimensions.

Units of physical character size include points, M-units and millimeters. Angular character size is typically measured in degrees or minutes of arc. In clinical applications, angular character size is often expressed using metrics from acuity testing including Snellen notation, logMAR, and decimal notation. Appendix A lists common definitions of character size and conversion formulas.

Second, sets of letters are often treated collectively as an ensemble of stimuli. Examples include the 10 Sloan letters used on the Early Treatment of Diabetic Retinopathy Study (ETDRS) acuity chart, or the 26 lowercase letters of a particular font, or the full set of upper- and lowercase letters of the font. The letters within the ensemble differ in shape and size. Given this variation, how is character size defined? A common convention, borrowed from typography, is to cite the size of a representative character, such as the height of a lowercase x (physical height or angular height). When this size is given, along with the identity of the font, the sizes of characters in the ensemble are fairly well defined.[2] If the ensemble consists of uppercase letters only, often true for acuity testing, all members of the ensemble have nearly the same height.

Sometimes, the size of letters for acuity testing is specified by the size of a "critical detail" (e.g., the gap in Landolt C stimuli or stroke width of the Sloan letters; National Research Council, 1980). In this case, the height of the corresponding test letters is several times greater than the size of the critical detail. For instance, for the 10 uppercase Sloan letters, character height is five times larger than the size of the critical detail.

---

[2] One subtlety in the designation of character size is that there is not a constant relationship between a font's point size and its x-height; the ratio depends on the font. Except for exotic fonts, however, one can expect the x-height of a 12-pt font to lie in the range from 4 to 7 pts (Charles Bigelow, personal communication, Sept. 12, 2004).

Third, when the horizontal center-to-center separation between letters in text is a pertinent measure, it is important to distinguish between *fixed-width* fonts, such as Courier, and proportionally spaced fonts (sometimes termed *variable-width fonts*) such as Times. In fixed-width fonts, each letter is allocated the same horizontal distance, regardless of the letter's width—an "i" occupies the same horizontal real estate as a "w." In a proportionally spaced font, the horizontal space is proportional to the width of the individual character. Analysis of the distribution of letter widths in a selection of common proportionally spaced fonts indicated that the standard deviation, expressed as a percentage of the mean character width, ranged from about 23% to 34% (Legge, Klitz, & Tjan, 1997). A major advantage of fixed-width fonts for vision testing is the direct proportionality between character spaces and visual angle. Put another way, any string of N letters from the same fixed-width font will subtend the same visual angle. For a proportionally spaced font, the angular width of a string will depend on the particular set of N letters in the string. An advantage of using proportionally spaced fonts in vision testing is their greater ecological validity; most everyday reading material is printed in proportionally spaced fonts. A practical advantage of proportionally spaced fonts is that more characters can be printed in the same horizontal line width. For instance, when matched for x-height, the mean horizontal space per character for Courier is 40% larger than for Times (R15, 1996).

Because of the direct correspondence between string length and angular character size, we have used fixed-width fonts in many of the studies in our series. In several of the early articles, we represented character size as the center-to-center separation of adjacent characters, equivalent to the horizontal real estate allocated to each letter in the fixed-width font.

3.2 THE EFFECT OF CHARACTER SIZE IN NORMAL AND LOW VISION

This variable is of dominant importance in many reading contexts including the size of print in newspapers and books, pixel resolution of electronic displays for text, legibility and salience of roadway and commercial signs, acuity testing, and the prescription of magnifiers for people with low vision. Several of the articles in our series deal with the effects of angular character size on reading, either for its own sake, or in relation to some other factor. Table 3.1 provides an overview.

Normal Vision

Speed-versus-size curves for normally sighted participants from several of our experiments are replotted in Figure 3.2.

The key pattern of results was first shown in two experiments reported in R1 (1985, Fig. 6) and replotted in Figure 3.2. The points show average reading speeds for four normally sighted participants as a function of character size for a 400:1 range from .06° (= 3.6 min-arc) to 24° (This size range is equivalent to seeing

TABLE 3.1

Articles in the Series Dealing With the Effect of Character Size

Papers and Figures	Subjects	Focus of Study	Method	Size Range (°)
R1, Figure 6	Normal vision	Character size per se	Drift	0.06–24
R2, Figure 2	Low vision	How do character size effects depend on the type of low vision?	Drift	Varies with subject
R5, Figure 3	Normal vision	Interaction with text contrast	Drift	0.13–12
R8, Figure 3	Normal vision	Compare character size effects with the drift and flashcard methods	Drift and Flashcard	0.06–24
R10, Figures 3 and 4	Normal vision	Interaction with age: Young and old subjects	Drift	0.15–12
R15, Figure 2	Normal and low vision	Interaction with font: Fixed-width and proportionally spaced	MNREAD chart	0.026–1.66[a]
R18, Figures 3 and 7	Normal vision	Central and peripheral vision	RSVP	0.06–0.36 in central vision

[a]The MNREAD chart has print sizes measured in logMAR units. For a standard reading distance of 40 cm, print sizes on the chart range from –0.5 to 1.3 logMAR (Snellen 20/6.3 to 20/400) corresponding to x-heights from .026 to 1.67°. The chart was designed so that no subjects are able to read the tiniest print in this range, ensuring no floor effects. Low-vision subjects in R15 were typically tested at viewing distances less than 40 cm, and correspondingly larger logMAR print sizes.

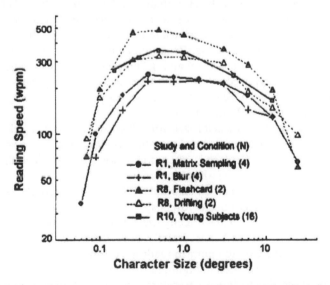

Figure 3.2   Reading speed versus character size. Data are replotted from five experiments in our series, as indicated in the legend. The data points are means across participants. (The matrix-sampling and blur experiments [R1, 1985] are discussed in section 4.3.)

6-inch letters from viewing distances ranging from 16 inches to about one tenth of a mile). The typical speed-versus-size curve shows a broad plateau of intermediate character sizes for which reading speed is fairly constant, a sharp decline in reading speeds for smaller characters, and a more gradual decline for very large characters.

There is a smallest print size below which reading speed begins to decline sharply, termed the *critical print size* (CPS). The CPS typically lies in the range from about 0.15° to 0.3° depending on the individual, stimulus factors such as font (R15, 1996), and the methods for measuring reading speed or for estimating the critical point. For instance, in R1 (1985) we estimated the CPS (the specific term was not used in the article) to be about 0.3°. In R18 (1998) the average CPS of six participants was .17° (range .14° to .24°). Across studies, a consensus value for the CPS for normally sighted readers is 0.2° (12 min-arc).

The concept of a CPS for reading has long been recognized, although the term is relatively new. Huey (1908/1968, chap. 21) recommended an x-height no less than 1.5 mm for printed texts. Huey's recommended x-height corresponds to an angular character size of 0.2° at a common reading distance of 40 cm. DeMarco and Massof (1997) surveyed the distribution of print sizes in 10 different sections of 100 U.S. newspapers. Median print sizes range from 0.78 M-units (stock listings) to 1.21 M-units (comic strips), corresponding to 0.17° to 0.25° for a 40 cm reading distance. (See Appendix A for the definition of M-units.) Evidently, newspaper designers know the CPS for reading.

It is important to distinguish between three related concepts: *CPS* for reading, *letter acuity,* and *reading acuity*. Letter acuity (often called Snellen acuity) is measured with short strings of unrelated test letters, and represents the smallest angular size for identifying the letters. Reading acuity refers to the measurement of visual acuity using a test chart containing sentences or words typeset as in text. Reading acuity is highly correlated with letter acuity. Some people with low vision, especially macular degeneration, have poorer reading acuity than letter acuity (Lovie-Kitchin & Bailey, 1981).

CPS is the smallest character size for which reading is possible at optimal speed. The CPS is at least two times larger than acuity letters for normally sighted participants, and the difference is often much larger for people with low vision (R2, 1985; Whittaker & Lovie-Kitchin, 1993).

The distinction between acuity size and CPS is important for the design of text displays and for the prescription of low-vision magnifiers. If a reading magnifier is prescribed to enlarge letters to the acuity limit, rather than to the CPS, the person's reading will be effortful and unnecessarily slow. Reading material should be enlarged to exceed the CPS to optimize reading performance. One contribution of our series of articles has been to highlight the importance of CPS for reading.

Reading speed also declines for very large characters, typically larger than 2° or 3° (R1, 1985; R10, 1991). To give an example of familiar Characters that subtend about 2°, the large digits on the corner of recently issued U.S. $20 currency bills are about 13 mm in height and subtend nearly 2° at a viewing distance of 40 cm.

There is roughly a 10-fold range of character sizes from 0.2° to 2° for which people with normal vision can achieve maximum reading speed.

What accounts for the inverted-bowl shape of the speed-versus-size curves? In R1 (1985), we mentioned three possibilities: limitations due to eye-movement control, some form of visual field restriction, and the spatial-frequency composition of letters of different sizes.

First, eye movements certainly impose an overall constraint on reading speed. As discussed in chapter 2, reading speeds measured with rapid serial visual presentation (RSVP), which reduces or eliminates the need for eye movements, are two to three times faster than reading with eye movements. Do oculomotor factors also explain the shape of the speed-versus-size curve? The large-print decline in reading speed was first documented in R1 (1985) where we used the drifting-text method. The decline in performance for large characters might occur because smooth-pursuit eye tracking has trouble keeping up with the high angular velocities associated with rapidly drifting large characters. Reduced reading speed for very small characters might also have an oculomotor origin. Kowler and Anton (1987) found that fixation times increase prior to short saccades (< 1°), presumably because of the demands of saccade planning. Although oculomotor factors may influence the details of the speed-versus-size curve, other factors must play a role because the qualitative shape of the curve is the same for RSVP reading.

Second, a form of visual-field restriction appears to have an important influence on speed-versus-size curves. We return to this issue in section 3.7 in connection with the visual-span model.

A third possibility is that reading speed is related to visual sensitivity for the spatial frequency composition of letters of different sizes. We return to this possibility in more detail in section 3.4 when we discuss the CSF model of reading.

Low Vision

The topic of character size is nearly synonymous with magnification in the context of low vision. Typical newsprint, held at 40 cm, is at the acuity limit of someone with Snellen acuity of 20/50. People with acuities poorer than 20/60, a common criterion for low vision, will be unable to read newsprint (and often many other forms of text) without magnification.

In R2 (1985, Fig. 2), we presented speed-versus-size curves for a heterogeneous group of 16 low-vision participants. Not surprisingly, most of the participants had CPSs much larger than normal.[3] Three conceptually important and clinically salient conclusions emerged from this work, extended and confirmed in later studies (R12,1992; R13, 1995).

---

[3]A notable exception was a subject with severe peripheral-field loss and a narrow bilateral wedge-shaped island of central vision. His Snellen acuity was 20/15. His speed-vs.-size curve (R2, 1985, Fig. 2B) was narrowly tuned with a peak near the normal critical print size of 0.2°.

First, some people with very poor acuity, requiring high magnification, can sometimes read with impressive speed. For example, a participant with Snellen acuity of 20/960 had a reading speed close to 100 wpm for white-on-black letters subtending 6° (R2, 1985, Fig. 2J), and a participant with 20/400 Snellen acuity had a reading speed of 206 wpm for 6° characters (R12, 1992). More generally, most people with low vision can achieve functionally useful reading speeds, given optimal viewing conditions, including magnification of character size to exceed their own CPS.

Second, letter acuity is only weakly correlated with maximum reading speed. In R12 (1992), we found that only 9.53% of the variance in reading speeds of 141 low-vision participants could be accounted for by letter acuity. Knowing a person's letter acuity says very little about their potential reading performance with suitably magnified text.

Third, the data in R2 (1985, Fig. 3) revealed an important dichotomy in low-vision participants. People with central-field loss (scotomas in the central part of the visual field including the fovea) usually read more slowly and required more magnification than people with at least some visual function within the central visual field.[4] This distinction is of broad significance to public health because the leading cause of impaired vision in developed countries is age-related macular degeneration (AMD), and AMD often results in central-field loss. The special reading difficulty of people with central-field loss was already known to clinicians (Faye, 1976), but our research brought this distinction into the laboratory and highlighted its importance for research.

People with central scotomas must use peripheral vision to read. How effective is eccentric vision for reading? This question, motivated by the problems of people with central scotomas, has been the focus of a great deal of recent research in our lab and elsewhere. We discuss this issue in section 3.6.

3.3 CONTRAST EFFECTS IN NORMAL AND LOW VISION

Because of the fundamental importance of contrast coding in vision, we investigated how changes in text contrast affect reading, and how deficits in contrast sensitivity can impair reading. Table 3.2 lists the articles in our series dealing with contrast and reading. After reviewing the findings in this section, we consider the interaction of contrast and character size in the next section.

Normal Vision

The basic findings for normal vision are presented in R5 (1987; Fig. 2) and R11 (1990, Fig. 3), and in Figure 3.3(a) of this chapter. For text contrasts, above a rather low critical value typically between .05 and 0.1, reading speed is nearly independ-

---

[4]In our series of papers, we often refer to subjects with functional vision within the central visual field as having "intact central vision." This does not imply that they have normal central vision, but only that they do not have an absolute loss of vision (absolute scotoma) within the central visual field.

TABLE 3.2

Articles in the Series Dealing With the Effect of Contrast Magnitude and Polarity

Papers and Figures	Subjects	Focus	Method
R1, Figure 6	Normal vision	Contrast polarity	Drift
R2, Figure 4	Low vision	Contrast polarity	Drift
R4, No figure	Normal and low vision	Polarity: Colored text or background	Drift
R5, Figures 2, 3, and 4	Normal vision	Magnitude and polarity	Drift
R6, Figures 1 and 4	Low vision	Magnitude and polarity	Drift
R11, Figures 3, 5, and 7	Normal and low vision	Luminance contrast and color contrast compared	Flashcard
R16, Figures 1–4	Normal vision	Magnitude: Effect on visual span and eye movements	RSVP and eye movements

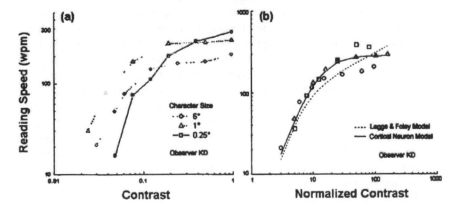

Figure 3.3 Reading speed versus luminance contrast. Reading speed is plotted as a function of contrast (a) for three character sizes and (b) after normalization by the contrast threshold for recognizing letters of the corresponding size. Data for three character sizes are shown for one participant. In (b), the dashed curve has the form of the compressive contrast-response function proposed by Legge and Foley (1980) to describe suprathreshold contrast coding. The solid curve has the form of a contrast-response function for a saturating cortical neuron. (Adapted from R5, 1987, Fig. 5.)

ent of contrast. This means that normal reading has at least a ten-fold tolerance for contrast reduction; people can read moderately low-contrast computer displays or faint Xerox copies with little impact on speed. When text contrast drops below the critical value, reading speed declines sharply.

There are some quantitative differences in the speed-versus-contrast curves for different angular character sizes (R5, 1987; R11, 1990) and for differences between text rendered with luminance contrast and color contrast (R11, 1990).[5] A simple scaling principle accounts for most of this variability. When text contrast (defined by luminance or color) is expressed as multiples of a threshold value, speed-versus-contrast curves largely superimpose across these stimulus variations (see R5, Fig. 5, and R11, Fig. 4). The scaling principle is illustrated in Figure 3.3. In panel (a), reading speed is plotted as a function of text contrast for one participant at three different character sizes—0.25°, 1°, and 6°. The curves have similar shapes but are displaced horizontally along the log contrast axis. In panel (b), the data have been replotted with text contrast expressed as multiples of the threshold contrast for letter recognition. When we factor out differences in contrast sensitivity at the 3 character sizes by normalizing in this way, the three sets of data superimpose up to a normalized contrast of 10. At higher levels there is some divergence.

What accounts for the shape of the speed-versus-contrast curves? Psychophysical models of contrast coding typically include a nonlinear transformation from stimulus contrast to visual response. The dashed curve in Figure 3.3 (b) is the compressive function used in the Legge and Foley model, which is derived from psychophysical contrast discrimination data (Legge & Foley, 1980). Contrast-response functions of simple and complex cells in cat and monkey visual cortex have a similar compressive form except that some of them saturate. The solid curve in Figure 3.3 (b) has the shape of a contrast-response function for a saturating cortical neuron. Both curves in Figure 3.3 (b) have the same qualitative features as the reading data, although the saturating neuron curve seems to provide a better quantitative fit. The point is that early stages of contrast coding, whether measured psychophysically or neurophysiologically, provide a plausible basis for contrast limitations in reading. Moreover, the scaling principle implies that the coding of contrast for reading is the same for characters of different sizes, and for rendering by color contrast or luminance contrast, apart from an early filtering stage that determines overall contrast sensitivity.

What is the effect of contrast polarity? Do people read better with white-on-black or black-on-white text? Westheimer (2003) showed that normal visual acuity is slightly better for reverse-contrast (white-on-black) Landolt rings (mean size ratio at the acuity limit for 4 participants was 1.41). Westheimer, Chu, Huang, Tran, and Dister (2003) also showed that the polarity difference increases with age. Although these findings might imply an advantage for white-on-black text for reading, some early findings, reviewed in R5 (1987), appear to show the opposite, a small advantage for black-on-white for reading or letter recognition. In our studies of normal reading speed for different character sizes and contrast levels, we have not found any systematic difference in reading speed for white-on-black and black-on-white text (R5, 1987, Fig. 2), or for black letters on a colored background and colored letters on a black background (R4, 1986). For normally sighted participants, we conclude that contrast polarity has little or no effect on reading speed.

---

[5]"Luminance contrast" refers to a difference in intensity between characters and background, whereas "color contrast" refers to a difference in chromaticity, e.g., red characters on a green background. See section 4.7 for more discussion of color effects on reading.

Low Vision

It is well known that some people with low vision read better with white-on-black text (Sloan, 1977). We compared low-vision reading speeds for the two polarities in R2 (1985), R6 (1989) and in a book chapter not included in the series (Legge, Rubin, & Schleske, 1986). The main finding is that people who read better with reversed-contrast text (white letters on a black background) usually have some cloudiness in the optics of the eye from cataracts, corneal damage, or vitreous debris. In R2 (1985, Fig. 4), we compiled the ratios of peak reading speeds for the two contrast polarities in groups of low-vision participants with clear and cloudy optics. For those with clear optics, the ratios were all very close to 1.0, indicating almost no difference in reading speed for the two contrast polarities. But all of the participants with cloudy optics read faster for white-on-black text, with the differences ranging from 10% to 52%.

What accounts for the contrast-polarity effect in people with cloudy optics? Opacities or irregularities in the optics of the eye produce light scatter, similar to viewing the world through a dirty windshield or spectacles smeared with petroleum jelly. The scattered light produces a veiling glare in the retinal image which acts like a whitewash to reduce retinal-image contrast of letters. The light-scatter effect is greater for black print on a white page because the stimulus field contains more white area than black. We found that for a typical white page of single-spaced black print, 84% of the page was white space and only 16% black. Legge et al. (1986) made detailed measurements on one participant with severe opacities and applied a quantitative version of the light-scatter model. They estimated that over 90% of the incident light at the participant's cornea was scattered into veiling luminance. The result was reduction in contrast sensitivity by more than a factor of ten and a strong contrast-polarity effect on reading speed: the participant read white-on-black text 44% faster than black-on-white text.

Contrast-polarity effects are not always associated with visible cloudiness of the eye's optics. Ehrlich (1987) found that 21 of 23 patients with severe forms of retinitis pigmentosa (RP) read faster with white-on-black text than with black-on-white text (mean ratio of reading speeds for the group was 1.26). Although cataract is frequently associated with RP, only 36% of Ehrlich's RP patients had detectable lens opacities. Subsequently, Alexander, Fishman, and Derlacki (1996) showed that abnormal levels of intraocular light scatter (measured with the van den Berg stray light meter) were present in a group of 20 RP patients, none of whom had more than trace lens opacities. The abnormal stray light, probably due to light scatter from morphological abnormalities in the crystalline lens, could explain the contrast-polarity effects observed by Ehrlich.

People who exhibit contrast-polarity effects in reading often benefit from electronic magnifiers that can produce contrast reversal. Closed-circuit television magnifiers have this feature. The flexibility of customizing color schemes in many computer applications is also helpful. For instance, some people with low vision choose to configure their Windows or Macintosh color schemes to routinely display text with bright letters on a dark background. Where possible, text displays for low vision should include an option for reversed contrast.

How does the magnitude of text contrast affect low-vision reading speed? Many low-vision participants, particularly those with cloudy optics, have a reduced contrast reserve, that is, the critical contrast at which reading begins to deteriorate occurs at much higher contrasts than normal (see R6, Fig. 1). In severe cases, reading performance declines for any reduction from maximum text contrast.

If the contrast-scaling principle for normal vision applies, then deficits in low-vision contrast sensitivity should be predictive of the impact of text-contrast reduction on reading speed. To address this issue, we measured speed-versus-contrast curves for 19 low-vision participants having a wide range of pathologies (R6, 1989). We used the data to study a generalization of the contrast-scaling principle. We termed this generalization to low vision the *contrast-attenuation model*. We evaluated two versions of the model. According to a strong version of the model, given suitably magnified print, any remaining deficits in low-vision reading speed can be ascribed to an effective reduction of text contrast. More specifically, low-vision speed-versus-contrast curves differ from the normal curve only by a contrast-attenuation factor (scale factor). This strong version of the model, implying that a deficit in contrast sensitivity is a primary factor limiting reading performance, provided a good description of the reading performance of a subset of the participants, those with cloudy optics. For them, once their reduced contrast sensitivity was taken into account by contrast-scaling the text, their speed-versus-contrast curves were essentially normal.

According to a weaker version of the model, the critical contrast for reading is predictable from a measure of contrast sensitivity, but even for text contrast above the critical value, reading speed is still depressed relative to normal performance. This version of the model implies that factors in addition to depressed contrast sensitivity impose a limit on reading performance. This weaker version of the model applied to participants with clear optics, typically with some form of retinal disease involving central vision. For them, contrast scaling of the text by an amount determined by their reduction from normal contrast sensitivity still left maximum reading speeds below normal values.

In summary, for people with low vision, the critical contrast for reading is often much higher than the normal value, with the difference predictable from standard measures of contrast sensitivity. For a subset of people with low vision—mostly those with cloudy optics—contrast deficits are a primary factor limiting reading performance. For many others with low vision, particularly those with central loss from retinal disease, reading speed is still depressed even when contrast is high enough not to be a limiting factor. We return to the reading problems of people with central loss in section 3.8.

3.4 THE CSF MODEL OF READING

In R5 (1987) we examined the interaction of contrast and character size on reading speed. We measured speed-versus-size curves for several contrast levels. Figure 3.4 replots data for one participant in this study.

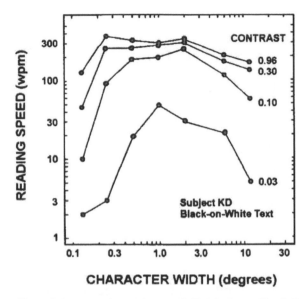

Figure 3.4 Effect of character size and contrast. Reading speed is plotted as a function of character size for four contrast levels. Data are shown for one participant. (Adapted from R5, 1987, Fig. 3).

Reducing text contrast first pushes down reading speed for very large and very small characters, and eventually, when the contrast gets very low, forces down performance for characters of intermediate size as well. The result is to produce a clear peak in the curve with high performance for middle-size characters and lower performance for small and large characters. This shape is reminiscent of the contrast-sensitivity function (CSF) for sinewave gratings,[6] and suggested to us that contrast sensitivity at different spatial frequencies might play a role in the character-size dependence of reading speed.

To explore this possibility, we created a CSF for reading (R5, 1987). Examples are shown in Figure 3.5.

The key steps in creating the CSF for reading were:

- For a given character size, we found the threshold contrast, arbitrarily defined as the contrast yielding a reading speed of 35 wpm.

---

[6]Building on linear-systems analysis, especially Fourier theory, early investigators used contrast thresholds for sine-wave gratings of different spatial frequency (bar widths of different sizes) to characterize human pattern vision (Campbell & Green, 1965; Campbell & Robson, 1968; Schade, 1956). The *contrast sensitivity function* (CSF)—a plot of contrast sensitivity (reciprocal of contrast threshold) as a function of spatial frequency—has become a standard tool for characterizing the impact of stimulus variables on human pattern vision and also the impact of eye disease. The CSF is characterized by a peak contrast sensitivity of about 300 at a medium spatial frequency of about 4 cycles per degree (cpd), a fairly rapid decline at high spatial frequencies to a cut-off spatial frequency between 30 and 60 cpd, and a more gradual decline at low spatial frequencies. Detailed characteristics of the CSF depend on stimulus properties such as luminance and temporal waveform.

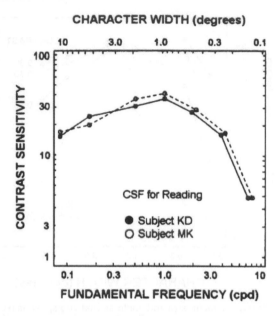

Figure 3.5   Contrast-sensitivity functions for reading. The contrast threshold for reading is defined to be the contrast required to read 35 wpm. Contrast sensitivity is the reciprocal of threshold contrast. The fundamental spatial frequency is equal to the reciprocal of the character width. Contrast sensitivities are plotted at twice the fundamental frequency of the character size. CSFs for reading are shown for two participants. The resulting curves are qualitatively similar in shape to CSFs for sine-wave gratings. See the text for more details. (From R5, 1987, Fig. 6).

- For each character size, we computed a fundamental spatial frequency of one cycle per character width, expressed in cycles per degree (cpd) by taking the reciprocal of the character width. For instance, a 2° character has a fundamental spatial frequency of 0.5 cpd.
- For characters of a given size, we plotted the contrast sensitivity at twice the fundamental spatial frequency. We chose this value because several studies of reading and letter recognition have shown that the most important information for letter recognition lies in the range of 2 to 3 cycles per letter (R1, 1985; Chung, Legge, & Tjan, 2002; Ginsburg, 1978; Majaj, Pelli, Kurshan, & Palomares, 2002; Parish & Sperling, 1991; J. A. Solomon & Pelli, 1994).
- To summarize, we created the CSF for reading as a graph of contrast sensitivity (reciprocal of threshold contrast for reading) versus twice the fundamental spatial frequency of the text letters.

We compared the CSFs for reading with a sine-wave grating CSF obtained under similar viewing conditions (R5, 1987, Fig. 7). The resulting CSFs for reading were very similar in shape to the corresponding sine-wave grating CSFs. These findings suggest that character-size effects in reading can be explained in terms of the spatiotemporal contrast sensitivity of vision. More specifically, the slow decline in reading speed for very large letters may be due to a corresponding decline in contrast sensitivity for low spatial frequencies. The more rapid decline in reading speed for very small letters may be associated with the steep fall-off in contrast sensitivity at high spatial frequencies.

In this section and the previous one, we have explained the effects of contrast and character size on reading speed in terms of existing models of contrast coding and contrast sensitivity in vision. We refer to this interpretation of our reading data as the *CSF Model* of reading.

Next, we briefly review two applications of this model. In R10 (1991), we compared reading speed for groups of young and old normally sighted participants (mean ages of 21.6 and 68.7 years) across a wide range of character sizes. We carefully screened our older participants to separate any with subclinical forms of cataract and other forms of eye disease from those with completely healthy eyes. Differences in reading speed between young participants and old participants with healthy eyes occurred only for very small and very large character sizes. We used the CSF model of reading to explain these age-related deficits in reading speed in terms of corresponding deficits in contrast sensitivity.

In another application, we evaluated the impact on contrast sensitivity and reading of multifocal intraocular lenses (Akutsu et al., 1992). Modern cataract surgery involves replacement of a patient's natural lens (crystalline lens) with an artificial intraocular lens (IOL). Most IOLs are fixed-focus, designed for clear distance vision. Reading text or other near work requires reading glasses for sharp focus. Designs have been proposed for multifocal IOLs (and also multifocal contact lenses) to increase the depth of field and reduce reliance on reading glasses. In our study, we compared sine-wave grating contrast sensitivities and reading performance for patients implanted with a multifocal IOL (diffractive optics design) and age-matched groups with normal vision or conventional monofocal IOLs. Group comparisons of the sine-wave grating CSFs confirmed the prediction from optics of an approximately two-fold reduction in contrast sensitivity for the multifocal patients. Application of the CSF model of reading predicted that the only impact on reading speed would occur for low-contrast text or for very small characters. This is precisely what we found (Akutsu et al., 1992).

These examples have strengthened our view that the CSF model of reading is a useful way to understand the interacting effects of character size and contrast on a participant's reading speed. The success of the CSF model of reading fuels the view that sensory mechanisms in vision influence reading performance.

3.5 THE SPATIAL-FREQUENCY CHANNEL MODEL OF READING

The CSF, measured with sine-wave gratings, is believed to be the envelope for more narrowly tuned sensitivity curves belonging to a set of spatial-frequency channels. These frequency-selective channels are thought to play a fundamental role in pattern vision (Blakemore & Campbell, 1969; Campbell & Robson, 1968).

Masking, pattern-adaptation, and other psychophysical methods have been used to estimate that these channels have a bandwidth between one and two octaves.[7] It is natural to ask what role these channels play in reading.

Ginsburg (1978) measured recognition for individual low-pass spatial-frequency-filtered letters. He found that letters required bandwidths between 1.5 and 3 cycles per letter for recognition. Assuming that not much useful information for recognition is present below one cycle per letter, Ginsburg's result implies a bandwidth of roughly a factor of two (one octave) of useful spatial-frequency information for letter recognition. He suggested that spatial-frequency information in a single channel is sufficient for letter recognition.

A priori, the bandwidth requirements for reading might be different from those for letter recognition. Since reading is a more complex task than letter recognition, involving segmenting of letter strings into words and execution of eye movements through text, reading might require a higher spatial-frequency bandwidth. It is also possible that low frequencies, below those useful for letter recognition, might be useful in coding coarse features such as word length or word shape. In R1 (1985), we measured reading speed for low-pass-filtered text, that is, for text with graded levels of blur. This experiment is reviewed in more detail in section 4.3, and examples of graded levels of blur are shown in Figure 4.3. In the experiment, decreasing bandwidth (increasing blur) had no effect on reading speed until a critical bandwidth was reached. For more severe blur, reading speed declined rapidly. The critical blur bandwidth was approximately 2.0 cycles per character, independent of character size.

In a subsequent report, not included in the series, we also measured reading speed for high-pass-filtered text in which frequencies below one cycle per character were removed (Beckmann, Legge, & Luebker, 1991). These low frequencies could potentially provide coarse word shape or word length information in reading. But removal of these low frequencies had no effect on reading speed, implying that coarse spatial-frequency features are not important to rapid reading.

Majaj, Liang, Martelli, Berger, and Pelli (2003) used a noise masking method to estimate the channel used in reading. They compared the properties of this channel to channels for letter recognition measured with the same method in a separate study (Majaj et al., 2002). The two types of channels were virtually identical, that is, the channel used for reading appears to be the same as the channel used for recognizing letters of the corresponding size. Their reading measurements showed no evidence for channels that might be using low frequency information for word length.

These three studies indicate that the important spatial frequencies for reading are to be found within a one- to two-octave range above one cycle per letter, a bandwidth consistent with processing by a single spatial-frequency channel. The pertinent channel would depend on character size. For instance, for text with let-

---

[7]The use of the strange unit "octave" as a unit of bandwidth was borrowed from psychoacoustics which in turn borrowed it from the musical interval. An octave refers to a factor of 2 in frequency. A channel bandwidth of 2 octaves would refer to frequency selectivity over a range of a factor of 4, e.g., 2 to 8 cycles per degree.

ters subtending 1°, a channel sensitive from 1 to 3 cpd might encode the information for reading. For letters subtending 0.2° (near the CPS), the relevant channel would be selective for frequencies from 5 to 15 cpd.

We do not yet have a consensus on the role of spatial-frequency channels in reading. In the remainder of this section, we review some complex findings addressing this issue.

The finding that the important spatial-frequency band for reading is equivalent to a channel bandwidth, and the supporting evidence for the CSF model of reading (section 3.4) together motivate the *channel model of reading*. In its strongest form, this model proposes that reading text of a given character size is dependent on spatial-frequency information restricted to a single spatial-frequency channel. The findings reviewed later in this section imply a weaker version of the model.

The channels mediating letter recognition have been studied in detail, and are undoubtedly relevant to reading. J. A. Solomon and Pelli (1994) used a noise-masking method to trace out the spatial-frequency tuning curves for channels mediating two very different tasks: identification of letters, and detection of sine-wave gratings. The tuning functions had identical shapes for the two tasks, implying that the same spatial-frequency channels were involved. As a specific example, J. A. Solomon and Pelli found that identification for 1° letters in the Bookman font was mediated by one spatial-frequency channel centered at 3 cycles per character. They believed that the center frequency for the relevant channel (expressed in units of cycles per letter) would remain constant across character sizes. This constancy is equivalent to saying that the center frequency of the channel, expressed in cpd, will scale in proportion to the inverse size of the letters—tinier letters are mediated by higher spatial-frequency channels.

Surprisingly, subsequent research in the same laboratory found deviation from linear size scaling (Majaj et al., 2002): the channel used for letter identification does not scale exactly with letter size. This effect is illustrated in Table 3.3.

This table is based on empirically derived formulas linking peak channel frequency to letter size.[8] Examples are shown for small letters (0.16°) and letters 100 times larger (16.6°). The peak frequency for the channel mediating identification of these letters can be expressed in units of cycles per letter or as cpd on the retina. Even though there is a 100-fold change in letter size in degrees, the peak frequency of the channel only changes by a factor of a little more than 20. The consequence of this lack of linear size scaling is that in units of cycles per letter, the peak frequency increases more than four-fold from 1.7 cycles/letter for the small characters to 7.7

---

[8]Majaj, Pelli, Kurshan, and Palomares (2002) found that letters of the same angular size from different fonts yielded somewhat different peak frequencies. They found that most of this variability could be eliminated by characterizing their letter fonts in terms of their *stroke frequency*. The stroke frequency for a font is the mean number of strokes cut by a horizontal line through the letters of the font. Majaj et al. plot their data using the stroke-frequency metric. Calculations in the table rely on three findings from Majaj et al. (2002): (a) The Bookman font has a stroke frequency of 1.7 strokes per letter. The numbers in the table would change for fonts with other stroke frequencies; (b) For letters with stroke frequencies of 10 strokes per letter, the corresponding peak channel frequency is 10 cpd; (c) Peak channel frequency (cycles per degree) increases with stroke frequency (strokes per degree) to the 2/3 power.

TABLE 3.3

Peak Channel Frequencies for Bookman Letters of Two Sizes (based on Majaj, Pelli, Kurshan, & Palomares, 2002)

Letter Size (°)	Channel Frequency (cyc/letter)	Channel Frequency (cyc/deg)
0.16	1.7	10
16.6	7.7	0.464

cycles/letter for the large characters. These empirical results imply that letters of large angular size are identified by channels encoding edge features or other higher-frequency components of the letters' spectra. Identification of tiny letters depends on channels that encode coarser features (lower frequencies in units of cycles per letter.)

The finding that the channel for letter recognition, designated by its peak frequency in cycles/letter, changes with character size is discrepant with the constant critical bandwidth of 2 cycles/letter (R1, 1985) discussed above. We return to a discussion of this discrepancy in connection with sampling density in section 4.3. One important and intriguing ramification of the departure from linear size scaling is that the nature of pattern recognition for letters depends on absolute angular size of letters (i.e., a different part of the letters spatial-frequency spectrum is selected for analysis, depending on the letter's angular size). This raises an important question for the channel model of reading. What dictates which channel is used for recognizing letters of a given size?

The CSF model for reading, described in the previous section, suggests one way of thinking about the size-dependent utilization of spatial-frequency information in letter recognition. Two factors contribute to determining the spatial-frequency band that is most informative for distinguishing among letters of a given size: the distinctiveness of information in different portions of the spectra of the letters themselves, and the degree to which the spatial-frequency information is attenuated by the participant's CSF. The impact of the CSF can be appreciated by thinking about tiny letters whose informative spectral components lie at high retinal spatial frequency, e.g., 0.1° letters for which 1 to 3 cycles per letter correspond to 10 to 30 cpd. These high frequencies are sharply attenuated by the rapidly descending high-frequency portion of the CSF, with contrast sensitivity much lower at 30 cpd than at 10 cpd. For these tiny letters, it would not be surprising if visual letter recognition relied more on information in the band 10 to 20 cpd than on information at higher spatial frequencies close to the high-frequency cutoff of the CSF.

These ideas were pursued in a study of contrast thresholds for bandpass-filtered letters (Chung et al., 2002). Our goal was to determine the band of spatial frequency in the spectra of letters of different sizes and retinal eccentricities for which the participant had highest sensitivity for identification. Adopting a method used previously by Alexander, Xie, and Derlacki (1994), we measured contrast thresholds for recognition of single letters (lowercase, Times Roman) of different sizes, presented in central and peripheral vision. The letters were filtered through nine narrow band-pass filters with peak spatial frequencies ranging from 0.63 to 10 cycles/letter, in half-octave steps. Plots of contrast sensitivity for letter identification versus peak frequency of the band-pass filters exhibited spatial-frequency tuning. Figure 3.6 (A) shows the resulting tuning curves for three sizes of letters in central vision. Letter size is expressed in log units relative to the size of acuity letters.

In agreement with Majaj et al. (2002), we found that the peak frequencies of these tuning functions shift to lower-frequency portions of the letter spectra for smaller letters; the curve for the 0.2 log-unit size is shifted leftward from the curve for the 0.6 log-unit size.

We were able to account for these changes in peak frequency across sizes and retinal eccentricity by a "CSF ideal-observer model." Figure 3.6 (b) shows data from the model. This model took into account the two key factors for identification mentioned earlier—the distinctive information in different portions of the letter spectra, and the weighting of this information by the participant's contrast-sensitivity function. More specifically, our ideal-observer model provides optimal recognition from image data, given knowledge of the spatial-frequency spectra of the set of possible letters, the presence of a front-end filter having the shape of the sine-wave grating CSF, and the addition of Gaussian white noise following the filter. The similarities of the model and the human data provide evidence that letter recognition is influenced by the psychophysical contrast-sensitivity function.

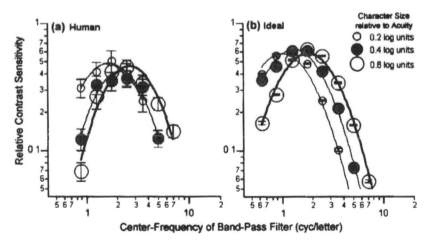

Figure 3.6   Contrast sensitivity for narrow frequency bands within the spectra of individual letters. Letters were filtered through narrow band-pass filters. Contrast sensitivity (the reciprocal of contrast threshold) for the filtered letters is plotted as a function of the center frequency of the filter. Data are shown for three character sizes, expressed in terms of log units relative to the size of acuity letters. (a) Human data. (b) CSF-ideal observer. These results are adapted from Chung, Legge, and Tjan (2002).

Our analysis indicates that participants use the most informative frequency band for letter recognition, and that this band changes with letter size because of the shape of the CSF. As a consequence, even before consideration of spatial-frequency channels in the visual pathway, size-dependent spatial-frequency tuning emerges (Fig. 3.6a) because of the combined influences of the shapes of the letter spectra and the CSF. In other words, even if the CSF represented the filter shape of a single broad-band channel, rather than the envelope of a set of narrow-band channels, we would expect to find relatively narrow-band spatial-frequency tuning for recognition of letters of a given size. This explanation does not deny the existence of spatial-frequency channels or their participation in letter recognition or reading, but asserts that frequency-selective tuning effects would be expected, even in the absence of these channels.

The claim that frequency-selective effects in letter recognition emerge from optimal use of spectral information rather than from underlying channels (Chung et al., 2002) is controversial. J. A. Solomon and Pelli (1994) and Majaj et al. (2002) interpret findings from their noise-masking studies of letter recognition to indicate that participants make an obligatory selection of a given channel, generally non-optimal in its information transfer, for letters of a given size. Their analyses do not explain why channel selection favors finer features (higher-frequency spectral components) for big letters, and coarser features for tiny letters.

Another tenet of the strong version of the channel theory of reading is challenged by additional findings from Beckmann et al. (1991). In that study, we constructed text in which some of the letters were filtered through one narrow band of spatial frequency and the other letters were filtered through a separate, non-overlapping band. The "low band" extended from 1 to 2.5 cycles/character and the "high band" from 4.5 to 12 cycles/character.[9] This wide separation between bands was intended to ensure that the low and high bands would stimulate different spatial-frequency channels. We compared reading speeds for three types of texts: all low-band characters, all high-band characters, and "ransom notes" in which the characters were randomly assigned to be low band or high band on a letter-by-letter basis. Consistent with the CSF model, for small letters, the low-band letters were easier to see and read than the high-band letters, because contrast sensitivity was higher for the frequencies comprising the low-band letters. For big letters, the reverse was true; the high-band letters were easier to read. For a character size for which pure low-band and pure high-band reading speeds were roughly matched, there was very little decrement in reading speeds for the corresponding ransom notes. If letters in text were processed by a single frequency-selective channel, sensitive to either the low band or the high band, we would expect that the mixed-frequency text would be difficult to read, since only about half of the letters would be recognizable with a single channel. Our contrary results imply that high-speed reading can be supported by sensory signals from widely separated spatial frequencies, inconsistent with the use of a single narrow-band channel in reading.

---

[9]The filters were 8th-order Butterworth filters.

This ransom-note result may be construed as evidence for parallel and independent recognition of letters in reading (more on the linkage of letters to reading in section 3.9). Perhaps it is true that channels tuned to different frequencies could operate in parallel to recognize neighboring letters in low and high bands in the ransom notes. In other words, single, narrowband channels might mediate recognition of individual letters in reading, even if multiple channels operate in parallel to handle the two kinds of band-pass letters in the ransom notes. Beckmann et al. (1991) performed a masking experiment to test the idea of frequency selectivity at the letter level. Participants were asked to read text rendered in the low band in the presence or absence of superimposed masking text in the high band. (They also did the reverse: target text in the high band and masker in the low band.) If letter recognition and reading rely on independent frequency-selective channels, we would expect that masking effects from text in a remote frequency band should be minimal. The logic is similar to masking studies with sine-wave grating detection.[10] Despite the wide separation between frequency bands in the text masking experiment, there were strong interference effects. There were no cases in which masked text in the low or high bands could be read at speeds close to the unmasked levels. These results imply that if single frequency-selective channels mediate early coding of information about letters, there is a commingling of information from channels tuned to different frequencies at or before the level of letter or word recognition. In other words, visual information for reading is not in general processed by a single channel, even at the level of letter recognition.

We are left with a weaker form of the channel model for reading. Yes, there is evidence that reading and letter recognition can proceed with bands of spatial frequency roughly matched in breadth to psychophysically defined channel bandwidths. In other words, the output of one channel could provide sufficient information for reading or letter recognition. But, in the reading process, there seems to be an obligatory analysis of letter features in different bands. This is useful to participants in the ransom note experiment, in which non-overlapping letters are filtered into different bands. It is a problem for participants in the masking experiment where they cannot selectively ignore features from masking letters based on their frequency content.

In section 3.9, we briefly return to the analysis of features in letters and letters in words. It may be the case that relatively simple features underlie the recognition of letters (Pelli, Burns, Farell, & Moore-Page, in press), perhaps encoded by feature detectors in early visual cortex (V1). These feature detectors may exhibit spatial-frequency and orientation selectivity. They may respond bottom-up, in a pre- attentive fashion, sending signals to a higher-level stage of letter recognition. Fast, bottom-up letter recognition may be promiscuous in accepting inputs from feature detectors with a wide range of coding properties including different spatial-frequency selectivity.

---

[10]For instance, Legge and Foley (1980) showed that threshold elevation for a 2 cycles per degree sine-wave grating in the presence of masker gratings diminished as the difference in spatial frequency between masker and target increased.

3.6 READING WITH PERIPHERAL VISION

The CSF model of reading explains contrast and character-size effects on reading in terms of early sensory coding. What about the effect of retinal eccentricity on reading performance? Performance on most psychophysical tasks deteriorates with increasing eccentricity in peripheral vision, but in some cases, performance becomes equivalent to central vision when stimulus size is appropriately scaled. Virsu and Rovamo (1979) argued that contrast sensitivity for sine-wave gratings can be equated across the visual field if the grating stimuli are scaled in size in accordance with the cortical magnification factor.[11] A straightforward extension of the CSF model of reading is to predict that an appropriate size-scaling of print size in peripheral vision should equate peripheral and central reading performance.

In addition to testing this scaling hypothesis, our interest in the reading performance of peripheral vision is motivated by relevance to low vision. People with central-field loss must use peripheral vision to read. We discuss the implications for low vision in section 3.8.

It is difficult to measure reading speed in the visual periphery of normal participants, because of their strong reflex to foveate the words. In R18 (1998), we used the RSVP method to measure peripheral reading speed. The RSVP method minimizes the need for eye movements because only one word is presented at a time (see section 2.1). In the study, RSVP sequences of words, comprising short sentences, were presented at 6 retinal eccentricities, from 0° to 20° in the lower visual field. The participants were not permitted to look down to fixate the words, although they were allowed to make horizontal eye movements along a fixation line.

For each of the six normal participants, we measured reading speed as a function of character size at each of the six eccentricities. All of the data followed the same typical pattern (R18, 1998; Fig. 3): for a range of small character sizes, reading speed increased with print size up to a CPS, and above the CPS, reading speed remained constant, independent of print size for the range tested. (In this study, we did not test at very large print sizes, greater than about 2°, for which we would expect a decline in reading speed as shown in Figure 3.2 of this chapter.) As would be expected from the well known decrease in spatial resolution, CPSs increased in peripheral vision, from 0.16° at the fovea to 2.22° at 20° eccentricity. This growth of CPS closely parallels the growth of single-letter acuity in peripheral vision (Herse & Bedell, 1989; Ludvigh, 1941), implying that both are limited by similar underlying neural factors.

The parameter E_2 is often used as an index for the rate of growth of threshold stimulus size in peripheral vision. It is the distance from fixation at which the threshold size doubles. The E_2 value for CPS was 1.4°, very close to the E_2 for single-letter acuity of 1.49° measured by Herse and Bedell (1989).

---

[11]Subsequent research has indicated that cortical magnification in visual area V1 has a steeper gradient than the scaling law for contrast sensitivity. Vernier acuity and other hyperacuities do follow a scaling law that is closely matched to cortical magnification (cf. Wilson, Levi, Maffei, Rovamo, & Devalois, 1990).

But inconsistent with models that equate central and peripheral vision through size scaling, the maximum reading speeds in peripheral vision never match those in central vision, even when character size exceeded the CPS. Across the six participants, average values of maximum reading speed decreased by about a factor of 6 from central vision to 20° eccentricity (807–135 wpm).

The failure of reading speed to follow a size-scaling model represents a departure from the straightforward extension of the CSF model of reading for peripheral vision. This departure is one result that has motivated us to consider an additional concept, the visual span, for explaining visual limitations on reading.

3.7 THE VISUAL SPAN

The visual span is the number of letters, arranged side-by-side as in text, that can be reliably recognized without moving the eyes. It is surprising to many people that the visual span is quite small. Panel A in Figure 3.7 illustrates the concept of the visual span, and why it is important for reading. Panel A shows that there is only a small region within the overall visual field in which text letters can be accurately recognized. For people with normal vision, this region is only about 10 letters wide. In other words, the normal visual span for reading is about 10 letters. Outside this region, the letters are too indistinct to be recognized reliably.

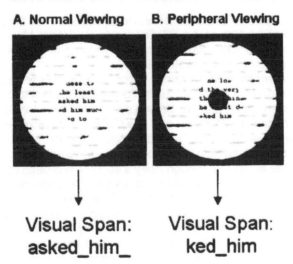

Figure 3.7   Schematic representation of the visual span. Panel A illustrates that in normal vision, there is only a small region of the visual field in which letters can be recognized accurately without moving the eyes. A reader can identify about 10 letters on a line of text passing through the center of this region, "asked_him_." This reader's visual span is said to be ten characters wide. Panel B shows that the region of reliable letter recognition becomes ring-shaped in the presence of a central scotoma, and that the maximum number of visible letters on a line of text is reduced. For example, if the reader attends to the line of text below the scotoma, the visual span is seven letters, "ked_him." The reader with a central scotoma has a smaller visual span.

As we discuss below, our research implies that the size of the visual span is determined by properties of early sensory coding. We propose that the size of the visual span is an important determinant of reading speed.

The modest size of the visual span is the primary reason why reading requires saccadic eye movements. Normal reading of printed text involves a series of fixations on a line of text, separated by saccades. The size of the visual span limits the number of letters that can be recognized in each fixation.

The notion that only a few letters are processed in each fixation in reading dates back to the 19th century (see Box 3.1). Our research has clarified the sensory origin of this limitation, and its relevance to reading. As discussed below, the impact on reading of stimulus variables including contrast, character size and retinal eccentricity can be understood through their effect on the size of the visual span. Similarly, important aspects of reading difficulty in low vision may be related to a reduction in the size of the visual span. For instance, Panel B in Figure 3.7 illustrates the impact of a central scotoma on the visual span, a topic to be discussed in section 3.8.

Box 3.1
Historical Antecedents of the Visual Span

The key idea that only a few letters are recognized at a time in reading dates back to the 19th century. Refinements of this idea have been introduced along with some subtle differences in terminology. Here is a brief summary of some of the milestones.

Javal, Lamare, and Hereing. According to Huey (1908/1968), Louis-Emil Javal, a French ophthalmologist, was the first to call attention to the discontinuous nature of reading eye movements. He distinguished between the pauses (fixations) and eye movements (saccades). According to Huey, Javal "concluded that there was a pause about every ten letters, and thought that this was about the amount that could be seen clearly at one fixation." Although Huey's account has been used to credit Javal with these discoveries, a recent historical analysis provides a new perspective (Tatler & Wade, 2003; Wade, Tatler, & Heller, 2003). Javal was certainly interested in the physiology of reading (he published a series of 8 articles on the topic in 1878 and 1879, cited by Tatler & Wade), and first used the term "saccade" to refer to rapid eye movements (Javal, 1879). But it was probably Lamare (1892), working in Javal's laboratory, and independently, Hering (1879) who first discovered that reading involved saccadic eye movements. According to Wade and Tatler, Hering detected reading saccades by an auditory method. By listening through two rubber tubes placed on the eyelids (a kind of miniature stethoscope), muscle contractions associated with eye movements could be heard, "one hears quite short, dull clapping sounds which follow each other at irregular intervals." The clapping sounds were present during reading, but not when subjects were instructed to fixate a static target.

Cattell (1886, 1885/1947). In two famous articles, employing ingenious methods for displaying stimuli and recording responses, Cattell showed that letters are recognized almost as quickly as words, implying a strong departure from serial letter recognition in reading. Cattell (1885/1947) used a drifting method (kymograph) so that 1, 2, or more letters were visible at a time in a viewing slit. The kymograph "consists essentially of a rotating cylinder driven by a clockwork at a speed which can be regulated at will." He showed that the recognition time per letter decreases as the number of letters visible in the slit increased, implying some overlap or parallelism in the recognition of letters. Cattell (1886) conducted a series of reaction-time experiments to tease apart the times required for sensory signals to reach conscious awareness, the additional time for a perceptual interpretation of the stimulus, and the motor response time. He developed a clever shutter system (gravity chronometer) for displaying stimuli, and both voice-activated and finger-activated switches for recording responses. In the letter recognition and word recognition experiments, the subject responded verbally or released a telegraph switch when a target stimulus was presented. He measured reaction times for responding to one target letter from a set of two, or from the full set of 26 letters of the alphabet,[i] and for identifying a target word from a set of 26 words. From the results, Cattell (1886) observed "It will be noticed that the perception-time is only slightly longer for a word than a single letter; we do not therefore perceive separately the letters of which a word is composed but the words as a whole." This provocative conclusion anticipated later research on word shape (discussed briefly in Section 3.9 of this chapter) and a great deal of debate among educators about methods for teaching reading.

Huey (1908/1968). Huey's book on *The Psychology and Pedagogy of Reading* was the first major effort to describe how perceptual and other cognitive factors influence reading and their relevance to reading instruction. He brought attention to the role of eye movements in reading and the small amount of information available on each fixation. He pointed out that people generally confuse the large amount of information that can be obtained from a page of print using eye movements with the small amount of information available in a single fixation. Huey introduced the concept of *reading range,* "if you will look fixedly at a letter in the middle of the page and will attempt to name the letters or words about it without moving the eyes for a single instant, you will discover that the reading range of the unmoved eye is distinctly limited" (pp. 51–52). By "reading range" Huey meant the distance from fixation at which observers could recognize words. He conducted experiments to estimate the size of the reading range. The apparatus was similar to Cattell's (1885/1947) kymograph. Huey presented his subjects with brief exposures ("peeps") to portions of

---

[i]Cattell (1886) observed substantial variation across the alphabet in the time required for recognizing different letters. There was very little difference in reaction times for uppercase and lowercase letters.

a sentence and asked them to read as much as they could. Knowing the fixation point, he could estimate the reading range as the number of letters to the right of fixation at which words could be recognized. He reported that some of his subjects could read more than 16 letter spaces to the right of fixation, but none could read words 26 letter spaces to the right of fixation. The subjects were allowed to read the preceding part of the sentence as well as the fragment of meaningful text in the slit, and so had the benefit of context. Huey recognized that the reading range depends on context and commented that it was smaller for isolated words. Anticipating future research, he identified three additional factors limiting the reading range: retinal factors associated with decreasing resolution in peripheral vision, attention to words in peripheral vision, and memory and other limitations associated with reporting what had been seen. Our trigram method for measuring visual span (Section 3.7) is intended to minimize effects of context, attention and reporting demands, while emphasizing the effects of front-end sensory factors on letter recognition.

Woodworth (1938). Woodworth addressed the issue of letter recognition in peripheral vision (he termed it "indirect vision") and its relevance to reading. By replotting data from previous investigators, he made clear that isolated letters can be recognized much farther into peripheral vision than letters flanked by other letters. In present terms, the visual span is much wider for isolated letters than letters within strings. Woodworth essentially describes our current notion of a visual-span profile: "Instead of a very narrow field of clear vision moving along the line and picking up one letter at a time, we have a broad field of fairly clear vision supplemented by a margin of less clear but still useful vision, this whole field advancing by jumps along the line" (p. 722).

Bouma (1970). Bouma's work on the impact of crowding on the visual span is discussed in Section 3.7. He used the term "functional visual field" to refer to the portion of the visual field around fixation within which letters could be identified. He quantified the crowding effect, showing that the functional visual field for letters flanked by other letters is three to four times smaller than the functional visual field for isolated letters. He further showed that the interfering effect of a flanking letter on recognition of a target extends over a rather large area, roughly equivalent to half of the distance of the target from fixation.

McConkie and Rayner (1975). These authors introduced the concept of "perceptual span" as the region around fixation in which printed information influences reading behavior. Operationally, it refers to the region of visual field that influences eye movements and fixation times in reading. McConkie and Rayner used a powerful new technique for estimating the size of the perceptual span. In the "moving window" technique, eye position is tracked while subjects read text on a computer display. The displayed text is updated on the screen, contingent on eye position, so that only letters within some designated distance of fixation, the "window," are visible, and text outside this "window" is masked. When the window is small, the pattern of eye movements is abnormal and reading slows down. When the window is very large, reading performance is normal. The assumption is that when the moving window is larger than the perceptual span, reading performance is unaffected by distortion of the text outside the window. This method was used to estimate that the perceptual span extends 15 characters to the right of fixation and leftward to the beginning of the currently fixated word up to a maximum of four characters (McConkie & Rayner, 1975; Rayner, Well, & Pollatsek, 1980). This large asymmetry of the perceptual span (15 characters right, 4 characters left) does not imply perceptual inferiority of the left visual field, but rather that stimulus factors influencing eye-movement control in reading extend farther to the right of fixation.

Our concept of "visual span" is quite different from McConkie and Rayner's (1975) concept of "perceptual span." Visual span characterizes letter recognition in the absence of oculomotor or contextual factors. Perceptual span is defined in terms of the functional demands of reading, including oculomotor demands and contextual effects. To illustrate two major differences between perceptual span and visual span: (a) There is only a slight left-right visual field asymmetry in the size of the visual span (R20, 2001) compared with the large asymmetry in the perceptual span; and (b) The visual span is defined and measured independent of reading context, but the size of the perceptual span depends on text difficulty (Rayner, 1986).

Rayner and Bertera (1979). These authors used a variant of the "moving window" technique to mask letters surrounding the point of fixation during reading. When the mask covered the central seven letters, reading speed was very low, about 12 words/minute. When the mask covered 11 letters, reading was essentially impossible. These results imply that human readers have a visual span of 7 to 11 letters.

O'Regan et al. (1983). These authors appear to have been the first to make empirical measurements very similar to our visual-span profiles. They measured the recognition of letters (flanked by numerals) as a function of their retinal eccentricity. They defined visual span in terms of the eccentricity within which letters could be recognized above some criterion level. For criteria of 50% and 90% correct, the visual spans were 22 letters and 10 letters respectively.

O'Regan (1990, 1991). He developed the first theoretical model for the shape of the visual span. According to the model, the number of adjacent letters recognizable in central vision is determined by the size of the critical features in the letters, the fall-off in the eye's spatial resolution away from the fixation point, and the geometry of the display surface. His model predicts a visual span of about 15 for letters subtending 0.4°.

R20 (2001). The last of the 20 articles in the Psychophysics of Reading series is the first to directly link properties of the visual span to reading speed. Given the very long history of development of the concept of visual span, it is surprising that it has taken so long to forge this explicit link.

Two Methods for Measuring the Visual Span

We have used two methods for measuring the size of the visual span. We prefer Method 2. Our first approach, Method 1, uses data from an RSVP word-reading task and some assumptions about how letter recognition participates in reading to infer the size of the visual span. Method 2 relies on a direct test of letter recognition, and does not involve measuring reading speed. Method 2 is better because it provides a direct assessment of the properties of the visual span which can then be related to other measurements of reading speed.

Before turning to Method 2 in detail, we briefly describe Method 1. It was used in some of our earlier work on the visual span, and we are not aware of any major inconsistencies in results obtained with the two methods.

Method 1: Reading speed versus word length.   In this procedure, a trial consisted of an RSVP sequence of random words of a fixed length. RSVP exposure times were varied to determine the threshold exposure time yielding 80% word-recognition accuracy. Testing yielded plots of threshold exposure time versus word length (see R20, 2001, Fig. 1).

The qualitative idea behind this method is that a narrower visual span should result in a stronger dependence of threshold exposure time on word length. If the visual span is very wide so that many letters can be recognized in one fixation, threshold should be insensitive to word length as long as word length is smaller than the size of the visual span. If the visual span is very narrow, threshold exposure times should rise for words whose length is greater than the visual span because it will take multiple fixations, separated by eye movements, to recognize all the letters of these words. As described in R16 (1997) and R20 (2001, Exp. 1), we developed a quantitative procedure for estimating the size of the visual span from the slope of regression lines fit to plots of threshold exposure time versus word length.

In Experiment 1 of R20 (2001), we used this method to estimate the size of visual spans from central vision to 15° in the lower visual field. Our goal was to distinguish between two alternatives. One possibility is that the visual span has the same size in central and peripheral vision, but word recognition slows down in peripheral vision, analogous to film that takes longer to develop. We termed this the "prolonged viewing" hypothesis. A second possibility, termed the "shrinking visual span hypothesis," is that visual spans get smaller in peripheral vision. The pattern of results clearly supported the shrinking visual-span hypothesis. From the results in Exp. 1 in R20, we estimated that the visual span shrinks from about 10 letters in central vision to about 1.7 letters at 15° in the lower visual field. The corresponding reduction in RSVP reading speed was 807 to 183 wpm (R18, 1998).

Although this method permitted us to estimate the size of the visual span in peripheral vision, we encountered two problems. First, the method only indirectly assesses letter-recognition accuracy, the key property of the visual span. Second, the data exhibited substantial individual differences. Although these differences might reflect real differences in perceptual coding, they might also be due to higher-level strategies, including eye-movement strategies for planning and executing saccades to stimuli in peripheral vision, or linguistic-inference strategies for "guessing" words from sparse visual data. These strategic influences are undesirable because we conceive of the visual span as a bottom-up sensory limitation on letter recognition. We developed a second method for measuring the visual span to circumvent these difficulties.

Method 2: The Trigram Procedure. In Experiment 2 of R20 (2001, we introduced a method for measuring the visual span that directly assesses letter recognition. This method is much less likely to be influenced by eye-movement strategies, or lexical inference.

The top panel of Figure 3.8 shows that our stimuli for measuring the visual span were trigrams, random strings of 3 letters. We used strings of letters rather than isolated letters because they include a key property of text—letters flanked on one or both sides by other letters. We measured performance for trigrams at different horizontal locations, with position indicated by the number of letter slots left or right of the midline. For instance, in Figure 3.8, the trigram "tgu" is positioned with "g" at slot 5. In central vision, position 0 corresponds to a letter at the point of fixation. For measurements in peripheral vision, e.g., at 10° in the lower visual field, position 0 corresponds to a letter on the midline 10° below fixation.

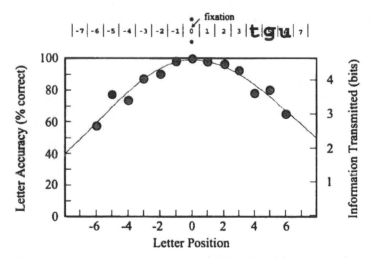

Figure 3.8 Measuring the visual span with the trigram method. Top: Illustrates that trials consist of the presentation of trigrams, random strings of three letters, at specified letter positions left and right of fixation. Bottom: Example of a *visual-span profile*, in which letter recognition accuracy (% correct) is plotted as a function of letter position for data accumulated across a block of trials. The right vertical scale shows the transformation from accuracy to information transmitted in bits (see text).

In a trial, a trigram was presented very briefly (e.g., 100 ms). The participant reported all 3 letters of the trigram. Across a block of trials, percent correct was accumulated for each letter slot.[12] We refer to the resulting plot of letter accuracy versus letter position as a "Visual-Span" profile, Figure 3.8 lower panel. These profiles usually peak at the midline and decline in the left and right visual fields. The profiles are often slightly broader on the right of the peak (R20, 2001). Similar asymmetries in the recognition of crowded letters have been reported by Bouma (1973) and Nazir, O'Regan, and Jacobs (1991). We have fit these profiles with "split Gaussians," that is, Gaussian curves that peak at letter position 0 but have unequal standard deviations on the left and right of the peak.

The right vertical scale for the visual-span profile in Figure 3.8 shows an approximately linear transformation from percent correct letter recognition to information transmitted in bits. The information values range from 0 bits for chance accuracy of 3.8% correct (the probability of correctly guessing one of 26 letters) to 4.7 bits for 100% accuracy. (For details of this transformation, see R20, 2001, Fig. 10 and associated footnote.) We quantify the size of the visual span by summing across the information transmitted in each slot (i.e. by computing the area under the visual-span profile in Fig. 3.8.). The 13 slots in the sample profile in Figure 3.8 transmit a total of 50.6 bits.[13] Lower or narrower visual span profiles will transmit fewer bits of information.

The visual-span profile represents the accuracy with which the visual system transmits letter information relevant for reading text. Think of the visual-span profile as a kind of information filter: within a single fixation, each letter of the stimulus word is "filtered" through the profile. The profile specifies the probability of correct recognition of each of the letters in the word. The further out on the tails of the profile, the greater the chance of a letter-recognition error.

In Experiment 2 of R20 (2001), we measured visual-span profiles for retinal eccentricities ranging from 0° (central vision) to 20° in the lower visual field. For each eccentricity, we used a print size that was larger than the CPS.

Figure 3.9 shows visual-span profiles at four retinal eccentricities. The curves are average values for three participants. As eccentricity increases, the profiles get narrower and drop below 100%. These results provide an independent confirmation of the shrinking visual span hypothesis described earlier. At 20° in the lower visual field, the peak of the visual-span profile is only 78%. This implies that reading at this retinal location would depend on unreliable letter recognition, with accuracies of 78% or less.

---

[12]This method for creating visual-span profiles accumulates letter-recognition accuracy in a given slot across cases in which the letter is the middle, left or right member of a trigram. Although recognition accuracy can be different for these three cases (R20, 2001, Fig. 5), reading of real texts includes all three cases for any given letter slot. By grouping data across these three cases in constructing visual-span profiles, we hope to represent performance across situations representative of reading.

[13]Since a letter slot with 100% accuracy transmits 4.7 bits, the total information transmitted through the sample visual span in Figure 3.8 is equivalent to 10.77 perfectly recognized letters.

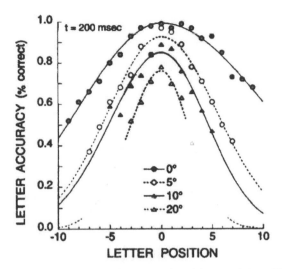

Figure 3.9   Effect of retinal eccentricity on the size of the visual span. Visual-span profiles are shown for four retinal eccentricities in the lower visual field. The data are averages for three participants. The exposure time was 200 msec. (Adapted from R20, 2001, Fig. 4).

Figure 3.10 is a scatter plot showing how the mean size of the visual span (measured as the amount of information transmitted in bits) covaries with the mean RSVP reading speeds at corresponding retinal eccentricities. The reading speeds were measured in R18 (1998) and the visual-span profiles were measured in R20 (2001). There is a very high correlation of 0.982, even though the visual spans and reading speeds were measured with different groups of participants. These results indicate a strong link between reading speed and the size of the visual span, and are consistent with the hypothesis that a shrinking visual span in peripheral vision is a cause of slower reading.

In R20, we also investigated temporal properties of the visual span, by measuring profiles for exposure times ranging from 25 msec to 500 msec. The profiles increased in size as exposure time increased (R20, Fig. 6). In central vision, the visual spans reached their peak amplitudes for exposure times of 100 msec or less. In peripheral vision, somewhat longer exposure times were required for visual spans to reach their peaks. This is evidence for slower letter recognition in peripheral vision. This central-peripheral difference in the speed of letter recognition may not have much impact on normal reading with eye movements. This is because mean fixation times are usually at least 200 ms, long enough for visual-span profiles to be maximized in either central or peripheral vision. In RSVP reading, however, where the exposure time per word can be arbitrarily short, the slower growth of the visual-span profiles in peripheral vision may contribute to slower reading.

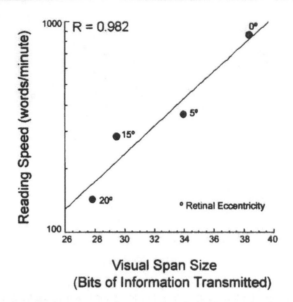

Figure 3.10   Relation between reading speed and visual-span size in peripheral vision. The figure shows a scatter plot of mean RSVP reading speeds from R18 (1998) and mean visual-span sizes from R20 (2001 at corresponding retinal eccentricities from 0° to 20° in the lower visual field. Visual spans were measured with 200 msec presentations. Visual-span size is given as the number of bits of information transmitted by the visual-span profile (see the text).

The Impact of Contrast and Character Size on the Visual Span

In earlier sections of this chapter, we examined how contrast and character size influence reading speed. If the size of the visual span imposes a sensory bottleneck on reading, we would expect to find that contrast and character size affect properties of the visual span in ways that are similar to their effects on reading speed. In ongoing experiments, we have investigated how visual-span profiles depend on character size and contrast (Legge, Lee, Owens, Cheung, & Chung, 2002).

Figure 3.11 shows visual-span profiles for five normally sighted participants tested at five contrast levels. A separate panel shows group data. At the lowest contrast (threshold contrast for letter recognition), the peak of the profile drops well below 100% accuracy. For higher contrasts, the peak rises to 100% and the profiles broaden; in other words, the visual spans get larger. For these same participants, we measured RSVP reading speed as a function of text contrast, obtaining curves similar to those shown in Figure 3.3. We found that the size of the visual-span profiles exhibited the same dependence on contrast as did reading speed.

The five panels in Figure 3.12 contain scatter plots showing how reading speed covaries with visual-span size for the five participants. The correlation coefficients

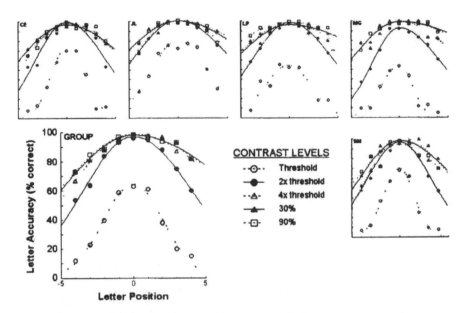

Figure 3.11 Visual-span profiles for different contrasts. Visual-span profiles (200 msec presentations) are shown for five contrast levels ranging from the contrast threshold for letter recognition (between 1% and 2% contrast) to 90% contrast. The five small panels show individual data for five participants. The large panel shows average results for the five participants.

are high, ranging from 0.95 to 0.99 across participants. Clearly, reading speed and size of the visual span are closely coupled in their dependence on contrast.

What is the impact of character size on the visual span? Figure 3.13 shows visual-span profiles for eight character sizes (left and middle panels) and the corresponding reading speeds (right panel) for one participant. The left panel shows profiles for four small character sizes, ranging from tiny letters of 3.78 min-arc (near this participant's acuity limit) to 10.6 min-arc. The profiles increase in height and breadth as character size increases, in correspondence with increasing reading speeds for these print sizes. The middle panel shows profiles for four larger print sizes, from 15 min-arc to 4°. For the largest print size, 4°, the visual-span profile diminishes in size, again corresponding to a decrease in reading speed for this participant and this character size.

It may seem paradoxical that visual spans actually get smaller for tiny letters. Should a visual span of a given size accommodate more tiny letters and fewer large letters? The resolution of this paradox is that the visual span is not a window of fixed size on the retina (or, equivalently fixed visual angle.) As discussed in detail in the next subsection, a consequence of the retina's decreasing spatial resolution

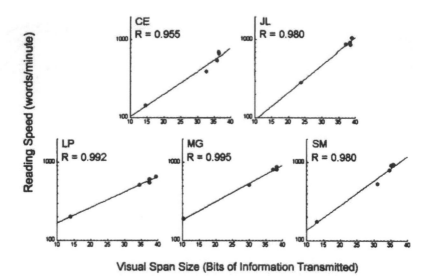

Figure 3.12   Relation between reading speed and visual-span size for different contrasts. Each panel shows a scatter plot of RSVP reading speeds and corresponding visual-span sizes for one of the participants in Figure 3.11. Each data point depicts the size of the visual span (bits of information transmitted) and corresponding RSVP reading speed for a given contrast level. The five points in each panel correspond to the five contrast levels In Figure 3.11. Regression lines have been fit to the data in each panel. The correlation coefficients range from 0.95 to 0.99.

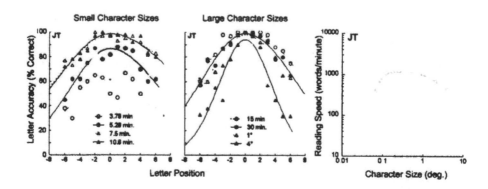

Figure 3.13   Visual-span profiles and reading speeds for different character sizes. Visual-span profiles are shown for one participant at eight character sizes ranging from 3.8 min (.06°) to 4°. The presentation times were 100 msec. A plot of RSVP reading speed at the same eight character sizes is shown for the same participant.

outward from the fovea is that the size of the visual span remains fairly constant across a moderate range of character sizes.

We have seen that three key stimulus variables cause reading speed and the size of the visual span to change in a highly correlated way—retinal eccentricity (Figs. 3.9 and 3.10), contrast (Figs. 3.11 and 3.12), and character size (Fig. 3.13). This close connection between size of the visual span and reading speed has motivated us to hypothesize that the size of the visual span is an important determinant of reading speed.

In R20 (2001) and in section 3.9 below, we discuss a computational model that implements this hypothesis. The model shows how empirically measured visual-span profiles have a direct impact on reading speed.

If size of the visual span were the only determinant of reading speed, then knowing the size of a person's visual span in bits would determine their reading speed in wpm, and scatterplots like those in Figs. 3.10 and 3.12 would superimpose. There are at least two reasons why this simple situation does not hold. First, there are individual differences in the relationship between reading speed and size of the visual span (see the individual curves in Fig. 3.11.) For a fixed amount of bottom-up information about letters in words, some people read faster than others. It is likely that these individual variations reflect differences in lexical access and the use of contextual or other linguistic knowledge (see section 2.2 for discussion of context effects in reading.) Second, as discussed earlier, the size of the visual span depends on exposure time, a relationship that changes in peripheral vision and possibly between participants. Clearly, the temporal dynamics of the visual span must be taken into account in assessing its effect on reading speed.

Three Determinants of the Size of the Visual Span

We now consider three mechanisms that influence the size of the visual span—decreasing letter acuity in peripheral vision, crowding between adjacent letters, and decreasing accuracy of position signals in peripheral vision.

Decreasing Letter Acuity in Peripheral Vision.   How does acuity for single letters change in peripheral vision? An important empirical finding is that the size of acuity letters grows linearly with retinal eccentricity out to at least 30° (Anstis, 1974; Weymouth, 1958) as described in the following equation:

$$S = S_0 + kE,$$

where S is the size of an acuity letter at retinal eccentricity E, S_0 is the size of acuity letters at fixation (where $E = 0$), and k is a constant. This equation is frequently rewritten in the following form:

$$S = S_0 (1 + E/E_2), \tag{1}$$

where $E_2 = S_0/k$ represents the eccentricity at which acuity letters double in size compared with their size at fixation. (Linear equations of this same form are used to express how the threshold size of many types of acuity—letter, word, vernier, grating, etc.—vary in peripheral vision, with the doubling constant E_2 indicating how rapidly the threshold changes with eccentricity.). This equation is plotted in Figure 3.14(a), for typical values of foveal letter acuity $S_0 = 5$ min-arc $(0.083°)$, and $E_2 = 1.5°$.

From this graph, we can estimate how wide the visual field is for isolated letters of a given size, that is, the distance from fixation within which they remain above the acuity threshold. For example, consider letters that are 10 min-arc in size, twice the size of acuity letters at fixation. The graph in Figure 3.14(a) shows that letters subtending 10 min-arc are at acuity threshold at an eccentricity of $1.5°$ (as expected, given that the E_2 parameter is $1.5°$). Beyond a distance of $1.5°$ from fixation, letters of this size will fall below the local acuity limit and will be unrecognizable. Because $1.5°$ is equivalent to 90 min-arc, we can say that the effective field size for 10 min-arc letters has a radius of 9 letters, or a diameter of 18 letters.

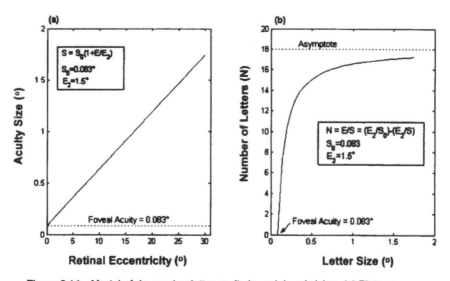

Figure 3.14   Model of decreasing letter acuity in peripheral vision. (a) Plots an equation showing how the size of acuity letters increases linearly with retinal eccentricity. The parameters of the linear equation are S_0, the size of acuity letters at the center of the fovea, and E_2, the retinal eccentricity at which the acuity size doubles. Typical values have been given to these parameters. (b) The equation from (a) is rearranged and plotted to show the distance into peripheral vision at which a character of a given size reaches its acuity limit, expressed in units of the number of letters N. The asymptotic value is 18. This means that relatively large characters fall below the local acuity limit at about 18 character positions from fixation. Small characters fall below the acuity limit at a smaller number of letter positions from fixation.

More generally, for any given letter size, we can compute the eccentricity E at which letters of that size reach the acuity threshold, and we can express this eccentricity in units of the character size. By doing so, we are expressing the eccentricity as the number N of letters of the given size that can be placed side by side from fixation. This is equivalent to rearranging Equation 1 to plot E/S as a function of S:

$$N = E/S = (E_z/S_0) - (E_z/S) \tag{2}$$

This equation is plotted in Figure 3.14(b). The curve asymptotes at a value of E_z/S_0 which, in our example, has a value of 18. This means that moderate size letters, substantially larger than foveal acuity letters, would be recognizable up to 18 letter positions outward from fixation. For letter sizes approaching the acuity limit, the curve drops rapidly. For instance, for letters just twice the size of foveal acuity letters (10 min-arc), we have already seen that the field size diminishes to nine letter positions from fixation.

When we use the metric N of letter positions, the approximate constancy of the field size for big letters, and the rapid decrease for tiny letters (approaching the foveal acuity limit), are direct consequences of the linear rule governing the change of spatial resolution for letters in peripheral vision (Fig. 3.14(a)).

If the decline in single-letter acuity in peripheral vision were the only limiting factor governing the size of the visual span for reading, we would expect:

- Because letters left and right of fixation are useful in reading, visual spans would be about twice the values shown in the graph in Figure 3.14(b), and could be as large as 36 letters.
- The number of letters in the visual span would be constant for letters substantially larger than the acuity limit.
- Near the acuity limit, visual spans would decrease in size.

The second and third points hold for visual spans measured with the trigram method, and implicate decreasing spatial resolution in the visual field as an important factor in determining the properties of the visual span. However, empirical measurements have shown that the actual visual span is far smaller than 36 characters, at most only a third or a quarter of this size. Other factors must limit the size of the visual span as measured with the trigram method.

Crowding.   It has long been known that recognition of letters, flanked on both sides (such as the letter "g" in the trigram "tgu") is much harder to recognize in peripheral vision than single letters (Bouma, 1970; Woodworth, 1938). This interfering effect of adjacent letters is commonly termed "crowding."

Bouma (1970) measured percent correct letter recognition as a function of retinal eccentricity for letters of a fixed size (the x-height was 14 min-arc). In separate conditions, target letters were presented alone, flanked on both sides with an "x" as in "xax," or flanked on one side only by an "x" as in "ax" or "xa." Recognition accuracy was severely reduced by the flankers and the interfering effect extended over a large distance from the target letter, roughly equivalent to half the distance from the target to the point of fixation. Surprisingly, the flanker farther from fixation than the target had a greater interfering effect than the flanker nearer fixation.[14]

Bouma (1970) estimated that the interfering effects of crowding reduced the functional visual field for letter recognition by a factor of four. If we apply Bouma's (1970) factor to the example above in which letter acuity alone predicts a visual span of 36, we arrive at a smaller value of 9, close to empirical estimates for the size of the visual span in normal reading. Bouma's (1970) findings strongly implicate crowding as a factor limiting the size of the visual span for reading.

Further evidence for the role of crowding in reading emerges from scaling laws in peripheral vision. The size of interference zones associated with crowding rise rapidly in peripheral vision with an E_2 value of about 0.7° (Levi, Klein, & Aitsebaomo, 1985), close to values for the hyperacuities. As mentioned earlier, acuity for single letters exhibits a more gradual change in peripheral vision with an E_2 value of about 1.5 (Herse & Bedell, 1989). But word-recognition acuity follows a scaling law with an estimated E_2 value of 0.68 (Abdelmour & Kalloniatis, 2001) very close to the value for crowding. This congruence of scaling laws forges a closer link between crowding and word recognition/reading, and suggests that the flanking effects on letter recognition are cortical in origin.

There is a puzzle underlying the presumptive links that connect crowding to reading speed. We would expect that stimulus factors that reduce crowding should increase the size of the visual span and also increase reading speed. We mention two counterexamples. First, crowding diminishes with increased spacing between letters. We would expect that increased letter spacing should result in faster reading, especially in peripheral vision where crowding is more pronounced. As discussed in section 4.2, extra-wide letter spacing does not increase reading speed in peripheral vision (Chung, 2002). A possible explanation for this unexpected result might be related to the effect of extra-wide spacing on the visual span. While adding extra space between letters may reduce crowding, the extra space also means that letters in the Nth slot away from the midline are pushed farther into peripheral vision. The joint effect might be to diminish the size of the visual span. Second, for some stimulus attributes, crowding diminishes when the flanking stimuli are dissimilar to the target, e.g. different contrast polarity or color (Kooi, Toet, Tripathy,

---

[14]An explanation for the interference associated with crowding has remained elusive. Some recent research has compared the properties of crowding to the interfering effects of "ordinary masking" in which a masking stimulus is superimposed spatially and temporally on a target (Chung, Levi, & Legge, 2001; Pelli, Palomares, & Majaj, 2004). Pelli et al. (2004) have presented a strong case for a distinct process underlying crowding that distinguishes it from ordinary masking. Whereas the spatial extent of crowding is proportional to the retinal eccentricity of the target but independent of size, the spatial extent of ordinary masking depends on target size and is fixed in spatial extent. Pelli et al. (2004) point out several other distinctions between crowding and ordinary masking, notably that crowding affects identification but not detection, while ordinary masking affects both identification and detection. Pelli et al. (2004) propose that crowding is due to feature integration over an inappropriately large "integration field" in peripheral vision.

& Levi, 1994). If reading benefits from reduced crowding, we might find that reading speed is faster for text composed of letters with alternating contrast polarity ("Oreo text" white-on-gray letters alternating with black-on-gray letters). But Chung and Mansfield (1999) found no difference in reading speeds between same-polarity and alternating polarity reading in central vision, or at 5° or 10° in the lower visual field. These examples illustrate that the impact of crowding on reading speed is still not well understood.

Position Signals.   The need for information about letter position distinguishes the recognition of words, from isolated letters. The strings "cat," "act," and "cta" differ only in the spatial order of their letters. Information about letter position must be encoded for proper lexical look-up. Our method for measuring visual-span profiles is sensitive to this positional information because a letter is scored as correct only if it is given in the proper position in the trigram.

Several researchers have considered the role of position signals in the recognition of letter strings including Estes (1978) and Mewhort, Campbell, Marchetti, and Campbell (1981). In a recent PhD thesis in our lab, Ortiz (2002) showed that an important part of the crowding of target letters by flanking letters in trigrams (or pentagrams) could be attributed to mislocation errors, that is, to the reporting of letters in the wrong spatial order. Ortiz estimated that mislocation errors accounted for about 50% of the slowing down of letter encoding and about 50% of the deterioration of identification accuracy. In other words, spatial uncertainty accounted for about half of the crowding of target letters by flanking letters. Cornelissen, Hansen, Gilchrist, et al. (1998) tested participants with two tasks requiring accurate encoding of letter position—a lexical-decision task and a primed reaction-time task. They found a correlation between performance on these tasks and thresholds for detecting motion coherence in moving random-dot patterns. Because thresholds for motion coherence are likely to depend on processing within the magnocellular pathway, Cornelissen, Hansen, Gilchrist, et al. (1998) proposed that letter-position information is carried by the magnocellular pathway.

To summarize, even under optimal conditions, visual spans for reading are quite narrow, extending only about four or five letter spaces left and right of the fixated letter. Three principal factors, related to increasing retinal eccentricity, contribute to this narrow window for letter recognition in reading: decreased spatial acuity, decreased accuracy of position signals, and increased crowding between adjacent letters.

3.8 CENTRAL-FIELD LOSS AND READING DIFFICULTY

Macular Degeneration and Central-Field Loss

There is an important clinical reason to study reading in peripheral vision. People with central scotomas from AMD or other forms of macular disease must use peripheral vision to read.

Macular degeneration is the leading cause of low vision in developed countries. A recent study estimated that there are 1.75 million people in the United States with AMD (Eye Diseases Prevalence Research Group, 2004). This number is projected to reach nearly 3 million by the year 2020. The prevalence of AMD rises sharply with increasing age above 65, and is much higher among White Americans than Black Americans.

The macula is a region of central retina, up to 5 mm in diameter (15°) characterized by yellowish pigmentation. It contains the fovea, the region of highest visual acuity, and the retinal reference point for eye fixations in normal participants. Behind the retina is a layer of light-absorbing cells called the retinal pigment epithelium (RPE). The RPE is separated from the blood vessels in the choroid (the layer between the retina and sclera) by Bruch's membrane.

There are two major forms of AMD—dry and wet. In the dry form, also known as geographic atrophy, there is patchy loss of RPE cells and damage to corresponding retinal photoreceptors. If the patchy loss does not include the fovea, acuity may remain high, and reading performance may be quite good. In the more serious wet form, sometimes termed choroidal neovascularization, new blood vessels from the choroid may penetrate through Bruch's membrane into the RPE. The vessels may rupture, releasing blood, and cause scarring that damages photoreceptors throughout the macular retina. AMD is often bilateral, affecting both eyes but not necessarily with the same time of onset or rate. At present, there is no prevention or cure for AMD. Given the high prevalence of AMD, the lack of a cure, and the adverse impact on reading (see below) and other activities of daily life, rehabilitation of people with AMD is an increasingly important public health issue.

Other types of eye disease can result in central-field loss. For example, there are several inherited forms of macular degeneration with onset in youth or early adulthood, including Stargardt's, Leber's and Best's disease. These are sometimes collectively termed juvenile macular degeneration (JMD).

Central-Field Loss Usually Means Slow Reading

Regardless of the underlying pathology, people with central loss usually have severe reading deficits (Faye, 1976; Fletcher, Schuchard, & Watson, 1999; Whittaker & Lovie-Kitchin, 1993; R2, 1985; R12, 1992).

In our first study of low-vision participants (R2, 1985), we made detailed measurements of reading speed versus character size (drifting-text method) on 18 eyes, seven with central-field loss and the remainder with intact fields or peripheral loss only. The maximum reading speeds for these participants showed a clear distinction: those with central loss read more slowly than the others. We summarized our findings in a simple regression model based on a 2 × 2 classification of ocular characteristics—(a) Did the participant have a central scotoma, yes or no? (b) Did the participant have cloudiness of the ocular media, yes or no? This four-way categorization accounted for 64% of the variance in maximum reading speeds. The predicted reading speeds from the regression model for the four groups illustrate the predominant influence of central-field loss:

- Intact central field and clear media, 132 wpm.
- Intact central field and cloudy media, 95 wpm.
- Central scotoma and clear media, 39 wpm.
- Central scotoma and cloudy media, 28 wpm.

Because the participant sample was small, and not representative of the distribution of ages and conditions in the overall low-vision population, it is likely that the mean reading speeds for the four groups would be different in another sample. Nevertheless, this was the first empirical study to quantify the reading difficulties of people with central-field loss, confirming clinical emphasis on the importance of this factor.

In R12 (1992), we reported measurements of reading speed on a larger sample of 141 low-vision participants. The flashcard method was used, and a very large print size (6° letters) ensured that character size exceeded the CPS for most of the participants. The mean reading speed of the 42 participants with no central scotomas was 112 wpm, compared to the mean of 215 wpm for an age-matched control group of normally sighted participants. There were 97 participants with central scotomas classified into three groups—41 with AMD,[15] 11 with various forms of JMD, and 45 with other causes of central-field loss. The reading speeds of the central-loss participants were quite variable (see R12, Fig. 6), and were significantly correlated with letter acuity and age. From regression models fit to the data, we can estimate that the reading speeds of central-loss participants with Snellen acuity of 20/200 (logMAR = 1.0) were, on averaged: AMD, 50 wpm; JMD, 94 wpm: Other central-loss participants, 100 wpm. These results confirmed our earlier finding (R2, 1985) that participants with central scotomas tend to read more slowly than low-vision participants with residual central vision. The variability of reading speeds among the central-loss groups made clear that a variety of factors modulate reading speed. In particular, we observed that acuity-matched JMD participants read about twice as fast as AMD participants.[16] We return briefly to a possible explanation for this age difference below in connection with the findings of Sunness, Applegate, Haselwood, and Rubin (1996).

We digress briefly to comment on the relationship between reading speed and visual acuity. The two studies just reviewed (R2, 1985; R12, 1992) and numerous others have shown that distance visual acuity for letters is only weakly correlated

---

[15]In R12, we used the alternative term age-related maculopathy (ARM) to refer to subjects with AMD.

[16]Lovie-Kitchen, Bowers, and Woods (2000) did not find a difference in reading speeds between groups of 13 AMD and 9 JMD subjects. They suggested that the discrepancy between their study and R12 might be related to the demands of the reading tasks. In their study, subjects were instructed to read for understanding and not to emphasize speed. In our study, subjects were pushed to their maximum reading speed.

with maximum low-vision reading speeds. For instance, in R12 (1992) the correlation between letter acuity and reading speed was statistically significant, but letter acuity accounted for only about 10% of the variance in reading speed. On the other hand, several studies have shown that low-vision reading speeds correlate more highly with measures of word acuity or reading acuity on tests such as the Sloan M cards (R2, 1985), Bailey-Lovie word acuity chart (Bullimore & Bailey, 1995), and MNREAD chart (Lovie-Kitchin, Bowers, & Woods, 2000).

Why Do People With Central Scotomas Read Slowly?

Central scotomas can have a wide range of sizes and shapes (see the review by Cheung & Legge, 2005). No doubt these variations contribute to the variability in reading performance across participants with central-field loss. Fletcher and Schuchard have used the scanning laser ophthalmoscope (SLO) to map the scotomas in large samples of low-vision patients (Fletcher, Schuchard, Livingstone, Crane, & Hu, 1994; Schuchard, Naseer, & de Castro, 1999).

Although the majority of central scotomas are dense, with complete loss of vision, up to 20% of central scotomas have spared islands of foveal vision (D. C. Fletcher, personal communication, n.d.). Sunness et al. (1999) reported that in a group of patients with geographic atrophy (the dry form of AMD), 8% had ring scotomas and 11% had horseshoe scotomas. The left panel of Figure 3.15 shows an example of a SLO map from an eye with a ring-shaped central scotoma. This patient has a small island of residual central vision, surrounded by a large region of macular scotoma. The right panel of Figure 3.15 shows a plot of reading speed versus character size (MNREAD test) for this patient. It is likely that the peak of the reading-speed graph occurs at a character size for which a small number of recognizable characters fits into the island of spared vision.

People with dense, central scotomas have no choice but to use peripheral vision for reading. An obvious problem is the decline in spatial acuity in peripheral vision. Decreasing acuity can be compensated for by magnifying print to exceed the acuity limit or, better yet, the CPS. As discussed in section 3.6, normal peripheral reading speed falls rapidly with increasing retinal eccentricity, even when the text characters exceed the CPS (R18, 1998). Paralleling this finding, participants with central-field loss benefit from magnification, but usually achieve maximum reading speeds well below normal levels (R2, 1985; R12, 1992).

What factors contribute to slow reading in AMD or other types of central-field loss? First, how plausible is the default assumption that peripheral vision in AMD participants functions like normal, healthy peripheral vision? Performance might actually be better in participants with central loss because of some form of long-term functional adaptation associated with use of peripheral retina for reading or other tasks. In support of this possibility, Casco, Campana, Grieco, Musetti, and Perrone (2003) reported that a 21-year-old patient with juvenile onset of macular degeneration (Stargardt's disease) out-performed normal controls

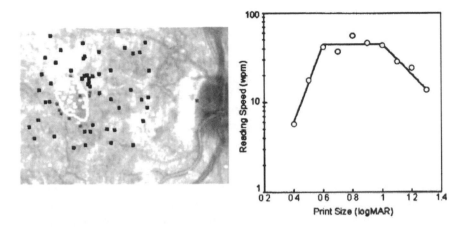

Figure 3.15 Visual-field and reading-speed measurements from a patient with a ring scotoma. A scanning laser ophthalmoscope (SLO) is a laser-based imaging device for viewing the retina. Some SLO's can also be used to present stimuli to observable locations on the retina, including stimuli for mapping scotomas. The left panel is a SLO visual field map. The white squares indicate retinal locations where the mapping target was reported seen. The black squares indicate retinal locations where the mapping target was not seen. An island of vision (cluster of white squares) is surrounded by scotomas (areas with black squares). This is an example of a ring scotoma. The right panel is a plot of reading speed versus print size (MNREAD test) from the same eye of the same patient. (Figure courtesy of Donald C. Fletcher.)

in a lexical-decision task (discriminating words from non-words) for stimuli at 5° away from the fovea. Alternatively, visual processing in peripheral retina of participants with central-field loss might be subnormal because of concommitant disease. For instance, Curcio, Owsley, and Jackson (2000) recently reviewed evidence that the normal age-related decline in scotopic sensitivity is accelerated in AMD patients, presumably related to deterioration of rod function. Unfortunately, we do not yet have comparative psychophysical data, including photopic acuity and other measures of letter identification, at matched retinal locations in normal and AMD eyes.

Even if peripheral vision is normal, we might expect eye movements to pose a problem. For effective reading. People with central scotomas must adopt some region of functioning peripheral retina for fixation and as a reference point for saccadic eye movements. There is evidence that it is difficult to adapt the saccadic system to use a nonfoveal retinal reference point (Peli, 1986; White & Bedell, 1990; Whittaker, Cummings & Swieson, 1991). If oculomotor recalibration is ineffective, reading would slow down. One way to minimize the impact of oculomotor factors is to use RSVP reading. In R18 (1998), we measured normal peripheral reading speed with RSVP. Speeds declined rapidly with increasing eccentricity, implying that some factors other than eye-movement limitations were at play. Rubin and Turano (1994) measured reading speeds for participants with central-field loss with RSVP and conventional static text. Their participants showed a small improvement in reading speed with RSVP compared with static text, but not nearly enough to restore normal reading speed. It appears that faulty eye-movement control is not the major limiting factor in the reading performance of people with central scotomas.

What is the impact of a central scotoma on the visual span for reading? Panel B of Figure 3.7 illustrates that the 2-D region of the visual field, normally available for reliable letter recognition, becomes ring-shaped in the presence of a central scotoma. It is no longer possible to recognize all the letters on a horizontal line of text passing through the center of this region. A person endeavoring to read with such a ring-shaped region could choose to attend to the line of text passing through the scotoma, with recognizable letters appearing in small clusters to the left and right of the scotoma. Alternatively, the reader could deploy attention to a line of text in peripheral vision above or below the scotoma. In either case, we would expect the number of recognizable letters (effective size of the visual span) to be reduced.

Suppose a patient with bilateral central scotomas is trained to read text in the lower visual field below her scotoma. How does her visual span compare with the normal visual span for reading in central vision? Assuming her peripheral vision is similar to normal peripheral vision, the findings from R20 (2001) reviewed above, imply that her visual span is reduced, with the extent of the reduction depending on the retinal location used for reading. In general, the greater the distance from the fovea, the smaller the visual span.

Even in a best case scenario in which a participant with central-field loss has healthy peripheral vision, uses the lower visual field for reading, has adequate magnification, and is not limited by eye-movement control, reading speed is likely to be reduced because of a smaller visual span. The reduced size of the visual span is a consequence of the anatomical structure of spatial vision.

Implications for Rehabilitation

The foregoing discussion seems to offer a rather dim prognosis for rehabilitation of reading in the presence of central scotomas. The reduced size of the visual span in peripheral vision imposes a fundamental sensory bottleneck on reading. But there is room for optimism in at least three areas.

First, "perceptual learning" refers to the capacity for improving perceptual performance through practice on perceptual tasks (cf. I. Fine & Jacobs, 2002; Sathian, 1998). This form of learning is presumed to be based on neural changes in the perceptual pathways, rather than on the learning of higher-level strategies for improving performance. Is the peripheral visual span modifiable in size through perceptual learning? If so, is there a corresponding improvement in reading performance? The answer to both of these questions is yes (Chung, Legge, & Cheung, 2004). In this study, the participants were 18 young, normally sighted adults. Testing was conducted at 10° in the upper and lower visual fields. Visual-span profiles and RSVP reading speeds were measured in pre- and posttests separated by several days. Between the pre- and posttests, participants received practice in four daily sessions (6 received practice in the upper visual field, 6 in the lower visual field, and 6 controls had no practice). The practice consisted of repeated measurement of five separate visual span profiles per day. The visual-span profiles of the trained participants continued to grow in size as practice proceeded. Of potential relevance to rehabilitation, the growth in size of the visual spans transferred to an improvement in reading speed. Post-test reading speeds were about 40% faster than pretest speeds for the trained participants, compared to an 8% improvement for controls who received no practice. The improvement in reading speed was even larger in a replication study (Lee, Gefroh, Legge, & Kwon, 2003).

These findings raise the possibility that perceptual learning, based on practice with letter recognition tasks like the trigram task, may be a useful rehabilitation strategy for people with central-field loss. Before this possibility becomes a reality, the utility of this type of training needs to be demonstrated with normal participants in the age range of AMD participants, and also directly with people with central-field loss.

A second issue relevant to rehabilitation is the selection of a peripheral retinal site for reading. Many people with central-field loss spontaneously adopt a nonfoveal retinal location for fixation and reading, often termed their preferred retinal locus (PRL). There is wide interest in the factors determining the selection of a retinal site for the PRL and the implications for reading. For a review, see Cheung and Legge (2005). It is generally believed that a PRL in the lower visual field below a scotoma (retinal location anatomically above the scotoma) is best for reading. Placement of the PRL above or below the central scotoma means that letters on a fixated line of text avoid the scotoma. Despite this functional advantage for reading, several studies reviewed by Cheung and Legge indicate that 30% to 60% of AMD patients spontaneously adopt a PRL to the left of their scotoma in the visual field. In Panel B of Figure 3.7, this is equivalent to using the portion of the visual span containing the letters "th" as the PRL. In such a case, upcoming letters on the line of text immediately to the right of the PRL will disappear into the scotoma. We would expect an adverse effect on reading speed.[17]

An important question for rehabilitation is to determine whether it is helpful to train patients with macular degeneration to use a more suitable peripheral retinal location for reading, sometimes termed "trained retinal locus." Nilsson, Frennesson, and Nilsson (1998) reported impressive improvements in reading

---

[17]Fine and Rubin (1999) tested normal subjects with simulated central scotomas. They demonstrated that a simulated scotoma to the right of fixation had a more adverse effect on reading performance than a simulated scotoma to the left.

speed when they trained a small group of AMD participants to use eccentric retina below their scotomas for reading.

Sunness et al. (1996) observed an interesting difference between AMD and JMD participants in site selection for the PRL. Unlike many of their AMD participants who placed the PRL to the left of the central scotoma in the visual field, most of the JMD participants placed the PRL below the scotoma. This difference in PRL placement would likely result in better reading performance for the JMD participants, And may explain the two-fold reading speed advantage for JMD over AMD participants we observed in R12 (1992).

Third, findings relevant to reading rehabilitation are likely to emerge from the use of neuroimaging to evaluate how visual areas of the brain respond to impaired vision. In the past 15 years, functional magnetic resonance imaging (fMRI), positron emission tomography (PET) and other neuroimaging methods have revealed a hitherto unexpected degree of plasticity in adult sensory systems. For reviews, see Buonomano (1998) and Kujala, Alho, and Naatanen (2000). Broadly speaking, three types of plasticity have been studied: (a) use-dependent expansion of the cortical representation, as in the fingers of string players (Elbert, Pantev, Wienbruch, Rockstroh, & Taub, 1995) or Braille readers (Pascual-Leone & Torres, 1993); (b) filling in of the cortical representation corresponding to a limited region of lost sensory function, as in finger amputation in adult monkeys (Merzenich et al., 1984) and bilateral foveal lesions in monkey retina (Heinen & Skavenski, 1991); and (c) cross-modality activation of cortical areas following long-term sensory deprivation, including tactile activation of visual cortex in blind participants (Buchel, Price, Frackowiak, & Friston, 1998; Sadato et al., 1996). There is even evidence that short-term visual deprivation—5 days of blindfolding—coupled with tactile training, can result in increased tactile activation of visual cortex and improved tactile discrimination in normally sighted people (Kauffman, Theoret, & Pascual-Leone, 2002; Pascual-Leone & Hamilton, 2001).

Most types of low vision, including macular degeneration, originate in diseases of the optics or retina. What is the relevance of brain-imaging studies of visual cortex? The recent evidence for brain plasticity makes it likely that visual cortical areas in people with low vision have adapted in some way. There are numerous important questions yet to be answered. Is the retinotopic map in V1 of people with central scotomas modified so that the large foveal projection is annexed for use by peripheral vision? Is the cortical representation of the PRL enlarged? Does the formation of the PRL involve a process of perceptual learning? What is the cortical site of the enlargement of the visual span from perceptual learning? Do people with low vision use visual cortex for tactile pattern recognition?

Therefore, despite the seemingly intractable limitations on reading in the presence of central scotomas due to shrinkage of the visual span, at least three lines of contemporary vision research may contribute to rehabilitation—perceptual learning, training in the use of an optimal retinal location for reading, and basic studies of cortical reorganization.

3.9 LINKING LETTER RECOGNITION TO READING SPEED

There is a large literature on psychophysical and perceptual issues in letter recognition, and now a growing literature on the psychophysics of reading. Although everyone would agree that letter recognition has something to do with reading, there is very little theory connecting the two. We have used the concept of the visual span to bridge the gap between letter recognition and reading speed.

Recall that visual-span profiles summarize letter-recognition accuracy in the portion of the visual field used for reading. Section 3.7 reviewed the evidence for a strong correlation between the size of the visual span and reading speed—both varied in a highly correlated way in response to changes in character size, contrast and retinal eccentricity. But how exactly does the size of the visual span influence reading speed? In the next subsection, we describe a model that predicts RSVP reading speeds given empirically determined visual-span profiles.

A Model Based on Letter Recognition

In R20 (2001), we presented a parameter-free model that explicitly links letter recognition to reading speed. The model uses the visual span as a bridging concept. The model is intended to explore the limitations on reading speed imposed by spatiotemporal constraints from letter recognition, rather than constraints from eye movements. For this reason, we designed the model to process RSVP text in which the effects of eye movements are minimized. This model is therefore different in style and purpose from several other recent models that explore the joint influences of word recognition and eye-movement control on reading. These include the strategy tactics model (O'Regan, 1990) the Mr. Chips model (Legge, Hooven et al., 2002; Legge et al. 1997;), the E-Z Reader model (Reichle, Pollatsek, Fisher, & Rayner, 1998), and the SWIFT model (Engbert, Longtin, & Kliegl, 2002).

The key features of the model are described with the help of Figures 11, 12, and 13 in R20 (2001). Figure 3.16, replotted from Figure 11 of R20 shows an example of the processing of the word "influence" by the model.

Here is a brief summary of the four basic parts of the model, with additional details to be found in the caption for Figure 3.16:

- The exposure time t of an RSVP word is allocated to one or more fixations.
- On each fixation, letter-recognition accuracy for each letter position in the word is determined by empirical visual-span profiles for the appropriate alignment on the word. Think of each letter in a fixation as being processed through an appropriate visual-span profile (filter); the further out on the tails of the profile, the greater the chance of a letter-recognition error (when errors occur, the model draws from a human letter-confusion matrix to determine the erroneous output letter). The narrower the visual span, or the lower its amplitude, the less information about the word's letters is transmitted through the visual span.

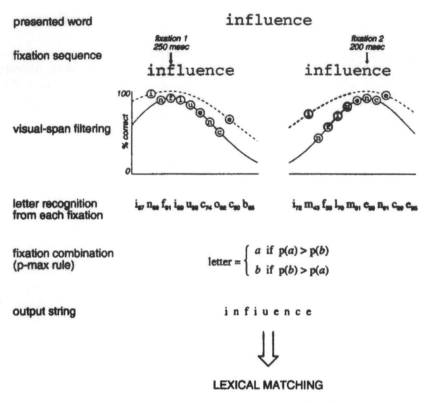

presented word · **influence**

fixation sequence

fixation 1
250 msec

influence

fixation 2
200 msec

influence

visual-span filtering

letter recognition
from each fixation

$i_{97}\ n_{98}\ f_{91}\ i_{80}\ u_{98}\ c_{74}\ o_{62}\ c_{50}\ b_{65}$

$i_{72}\ m_{43}\ f_{93}\ l_{70}\ m_{91}\ e_{98}\ n_{91}\ c_{99}\ e_{95}$

fixation combination
(p-max rule)

$$\text{letter} = \begin{cases} a \ \text{if} \ p(a) > p(b) \\ b \ \text{if} \ p(b) > p(a) \end{cases}$$

output string

i n f i u e n c e

⇩

LEXICAL MATCHING

Figure 3.16   Example of how the model encodes letter information from an input word. Here, the input word is "influence" presented at 5° eccentricity for 500 msec. The model makes two fixations, lasting 250 and 200 ms, centered on *f* and *n*. (The intervening saccade is assumed to take 50 msec.) Letter-recognition accuracy is determined by empirical visual-span profiles for the appropriate retinal eccentricity and for presentation times matching the fixation durations. The parameters of these profiles are taken from data summarized in R20 (2001, Fig. 7). For the first fixation, we use a profile for a 250-ms presentation time and *middle* letters to determine recognition accuracy for the interior seven letters of the input word. Accuracy for the end letters is determined by a 250-ms profile for *outer* letters. In the diagram, the output letters from a fixation are subscripted with the accuracy of the corresponding slot on the visual-span profile. In the first fixation, for example, *c* is correctly transmitted through a slot with only 50% accuracy, but one of the *es* is incorrectly transmitted as *c* through a slot with 74% accuracy. Because transmission errors occur, the output strings from the two fixations are not identical. Discrepancies are resolved by the *p-max* rule: decide in favor of the slot with higher accuracy. This rule leads to a correct selection of *n* over *m* for the second letter of the word, but an incorrect selection of *l* over *l* for the fourth letter. The final output string "influence" contains one transmission error. A lexical matching rule is then used (see text) to match this letter string to known words in the lexicon. The properties of this model are summarized in the present chapter, and discussed in detail in R20 (2001).

- The fixations on a word generates an output string of letters which may contain errors. If there are no errors, the model's output letter string matches the stimulus word; otherwise, there is a mismatch. There are no free parameters in the model, but we do have to specify rules for matching the output string to words in the model's lexicon (dictionary). In Reading 20, we discussed two bounding rules for lexical matching: *Exact Matching and Ideal Matching.*

The Exact Matching rule requires an Exact Match between the output letter string and a word in the lexicon before the model can read the word. If any output letter is in error the model fails to identify the word. This version of the model has no capacity for lexical inference; any deviation from an exact match and it misses the word.

The ideal matching rule puts an upper bound on the reading speed achievable through lexical inference, given the letter-recognition information transmitted by the visual span. This rule uses Bayesian inference to compute the most probable input word, given the output letter string, knowledge of the letter confusion matrices and word frequencies.

The version of the model with Exact Matching provides a good fit to the human data in central vision (R20, 2001, Fig. 13). This good fit means that the bottom-up visual data transmitted by the visual span, together with a very simplistic lexical matching procedure can account for high-speed RSVP reading in central vision.

We also compared the reading speeds of the Exact-Matching version of the model with human RSVP reading speeds in peripheral vision data from R18 (1998). In this case, the model's reading speeds drop off more rapidly than human reading speeds in peripheral vision. Unlike the model, people apparently do some form of lexical error correction on the fly when the bottom-up data from the visual span become less reliable in peripheral vision.

The version of the model with the Ideal Matching rule easily outperforms humans in both central and peripheral vision (R20, 2001, Fig. 13). At 20° in peripheral vision, its reading speed is about equal to human reading speed in central vision. This shows that, in principle, the rather inaccurate information about letter identity transmitted by the visual span at 20° eccentricity can, in theory, support quite rapid reading.

Because human reading performance in peripheral vision exceeds the lower bound from Exact Lexical Matching, but falls far short of Ideal Lexical Matching, we conclude that humans use some non-optimal form of lexical error correction when the bottom-up visual data about letter identity becomes unreliable.

What sort of mechanisms might people use for lexical error correction in peripheral reading? We briefly comment on three possibilities. One possibility is that knowledge about the visual confusion matrix (i.e., visual similarity) is used in error correction. For instance, since "e" is more often confused with "c" than "r," the nonword "aet" is more likely to be "act" than "art." A second possibility is that sentence context is used in error correction. As discussed in section 2.2, reading speed is faster for sentences than for meaningless strings of words, presumably because the syntactic and semantic relationships between words supplement bottom-up letter information in reading. Syntactic and semantic constraints could also be used to resolve uncertainty about words when information about letter identity becomes unreliable. Context effects, however, are not large enough to explain the entire gap between human peripheral reading speed and performance expected from exact lexical matching. In R18 (1998), we compared RSVP reading speeds in peripheral vision for sentences and meaningless strings of words. The reading speeds for the meaningless strings were still substantially faster than the speeds predicted from Exact Lexical Matching. A third likely possibility is that people use lexical inference (knowledge about words per se) to figure out words with letter-identity errors. For instance, the string "caterpiqlar" would not be recognized by Exact Lexical Matching, but most human readers would infer the intended word to be "caterpillar."

The insights from this model encourage us to hope that a computational framework, based on empirical visual-span profiles, can be used to understand the linkage between letter recognition and reading speed.

We have adopted the theoretical view that letter recognition precedes word recognition in reading, and is fundamental to it. We have taken this stance on the grounds of parsimony, recognizing that there is a long and unresolved debate about the perceptual units in reading (letters, words, or something else). Our findings indicate that a simple, bottom-up (letters first) model accounts for RSVP reading speed in central vision. But what about the idea that words are recognized directly by their shape, rather than by means of their letters?

Word Shape

Some people have the compelling intuition that they use the shapes of words to recognize them, bypassing recognition of the individual letters. In some special cases, this is undoubtedly true, as in recognition of words designated by special symbols rather than letters (e.g., the symbol "&" for the word "and"). However, most instances of word recognition in reading appear to rely on letter recognition. Here are four lines of evidence speaking to this issue.

1. A common proposal for the definition of word shape is the pattern of ascenders and descenders in the lowercase versions of words. For instance, "dog" has the pattern Ascender-Neutral-Descender. Since this form of word shape information is absent in uppercase words (compare "stopped" and "STOPPED"), we would expect word shape to enhance reading speed for lowercase compared to uppercase text. The empirical findings are mixed. Tinker (1963, chap. 4) reviewed several studies from which he concluded that there is indeed a decided speed advantage for lowercase text. However, F. Smith (1969) and Arditi and Cho (2000b) found no significant difference.

Walker (1987) analyzed the three- to seven-letter words in the Kucera and Francis (1967) corpus to determine how diagnostic the patterns of ascenders, descenders and neutral letters are for word identity. He found that only rarely do words have unique word shape. For instance, just two out of 439 three-letter words have unique word shape—"bye" and "gyp." Combining word shape with identity of the initial letter of words increases the proportion of words uniquely specified to between 3.6% and 13.6% depending on word length. When initial and final letters are known, along with word shape, between 17 and 33% of words are uniquely determined. Walker's analysis makes clear that word shape (defined as the pattern of ascenders and descenders) is, at best, a supplementary cue to letter identity in word recognition.

2. An alternative view of word shape is that important information about word identity is contained in the coarse spatial structure of words, conveyed by spatial frequencies below those that are important for letter recognition. Word length is an example of information that might be carried by low spatial frequencies. To determine if these low frequencies facilitate reading, we compared speeds (flashcard method) for unfiltered text and text that was highpass-filtered to remove spatial frequencies below one cycle per letter (Beckmann et al., 1991). There was no significant difference in reading speed for the filtered and unfiltered text. This result means that coarse information about word shape, carried by spatial frequencies below those used for letter recognition does not influence reading speed.

3. If words are analyzed as unitary shapes, then factors that disrupt pattern grouping should disrupt word recognition and reading. It is known that figural grouping is influenced by contrast polarity. For example, the white squares on a chess board may group perceptually to give the impression of diagonals. Interleaving light and dark letters within words would be expected to disrupt word shape and reading. But two studies agree in showing that reading speed for mixed-polarity text is no slower than reading speed for text composed entirely of letters of one polarity (Beckmann et al., 1991; Chung & Mansfield, 1999).

4. The foregoing studies cast doubt on the importance of coarse shape cues of words such as the bounding contour. It remains possible, however, that all of the contours of a word comprise a complex shape that is recognized as a single object. One variant of this idea would be a template-matching system in which there are separate templates for each word rather than for each letter. Pelli et al. (2003) performed a clever test of this idea. They asked, "Is a familiar word recognized as an image, or as a collection of letters?" They measured statistical efficiency for recognizing low-contrast letters or words in luminance noise. Under such conditions, best possible performance ("ideal-observer" performance) is limited by the signal-to-noise ratio (where "signal" is determined by the contrast energy of the stimulus). Typically, human participants require more stimulus contrast (that is, a higher signal-to-noise ratio) to achieve a criterion level of performance than the ideal observer. The human participant's efficiency is the (squared) ratio of human contrast threshold to ideal threshold. If humans recognize words as wholes, we would expect efficiency for word recognition to match efficiency for letter recognition. A consequence is that contrast thresholds in

noise should be lower for words than individual letters (since contrast energy is integrated across the word). Inconsistent with this expectation, Pelli et al. (2003) found that contrast thresholds are no lower for words, and that efficiency for recognition of words is inversely proportional to word length. This pattern of results is what would be expected if word-recognition performance is limited by efficiency for recognizing individual letters, rather than individual words.

Pelli et al.'s (2003) findings provide strong evidence against the proposal that word images are analyzed as whole visual patterns. The alternative view that emerges is that letters are recognized first, and then some process assembles letter identities into word recognition. This conception of letter and word recognition is consistent with the model linking letter recognition to reading speed presented in R20 (2001) and discussed earlier in this section.

Before leaving the topic of word shape, be aware of a potential source of confusion between the concepts of "word shape" and "word form." Word shape refers to some spatial shape property of words that can be defined on the image representation of words. Word form refers to a higher level representation of the arrangement and identity of letters, in which information about image features is no longer retained. It has been hypothesized that there is a left-hemisphere region in the brain dedicated to word form. Evidence for this "visual word form area" is briefly reviewed in Box 3.2.

Box 3.2
Is There a Visual Word Form Area?

Where in the stream of information processing in the brain does the transformation take place from letters and words as visual features to nonvisual representations? There is evidence from neuropsychological[1] and brain imaging studies that this transformation occurs in a left hemispheric ventral region near the boundary of the occipital and temporal lobes. More specifically, the left fusiform gyrus. This region has been termed the "visual word form area" or VWFA (L. Cohen et al., 2000; L. Cohen et al., 2002). It is currently controversial whether this region is specifically dedicated to "word form" analysis in reading. This controversy is reminiscent of the debate over the degree of specialization of the fusiform face area for analyzing face stimuli.

The notion of "word form" is distinct from the notion of "word shape" (section 3.9) and refers to a representation that encodes the identity and spatial arrangement of letters, but not visual features conveying information about font, color, size, upper/lower case, or other image features. For example, the same word-form representation would be derived from

---

[1]The notion of "word form" was introduced by Warrington and Shallice (1980) to describe an early stage of purely visual analysis of written words. They were concerned with a form of acquired dyslexia in which "word form" analysis appears to be impaired.

the following stimulus words—elephant, *elephant,* ELEPHANT, **elephant.** "Word form" does not imply word meaning or even that the string
has been recognized as a familiar word; Word form is thought to be a
representation that precedes contact with a mental lexicon or phonological representation. Word-form representations could also include
strings of letters composing nonwords.

There is both theoretical and empirical support for the existence of a
word-form representation. Such a representation is implicit or explicit in
most bottom-up models of word recognition or reading in which letter
recognition precedes word recognition. In the model described in R20
(2001, also reviewed in section 3.9), the output string from the visual
span is a word-form representation. The dual-route model (Box 3.3)
could include word-form representations at the front end of both the lexical and non-lexical routes. Empirical support for a word-form representation comes from studies indicating that pre-lexical effects of letter
strings can be invariant to changes between upper and lower case. For
instance, facilitation due to parafoveal preview survives a change in case
(upper to lower) across fixations in reading (Rayner, McConkie, & Zola,
1980). Similarly, priming effects in word naming are invariant to case
changes from upper to lower (e.g., see J. S. Bowers, Vigliocco, & Haan,
1998).

Brain-imaging studies may provide another source of evidence for a
word-form representation and may localize the visual word-form area in
the brain. Cohen et al. (2002) used functional magnetic resonance imaging (fMRI) to identify a left inferotemporal area as the VWFA. This region
showed stronger activation for alphabetic strings than checkerboard
patterns, stronger activation for words than strings of consonants, but
equal activation for words and pronounceable non-words. Dehaene et
al. (2001) used an fMRI priming method (suppression of response to repeated stimuli) to identify a brain region in the left fusiform gyrus that exhibits case-invariant word priming. Dehaene, Le Clec, Poline, Bihan, and
Cohen (2002) obtained evidence that this VWFA is modality specific, that
is, responds much more to visual than spoken stimuli.

The claim that the VWFA is specialized for visual words has been challenged by Price and Devlin (2003). They assert that there are no known
neurological cases of problems limited to visual word form and lesions
confined to the left fusiform area. They cite evidence indicating the VWFA
is activated in normal subjects by tasks other than visual word-form processing including naming pictures or colors, and repeating auditory
words. In short, Price and Devlin contend that the VWFA is a myth. Even if
this brain region plays a role in visual word analysis, they warn that we
should not conclude that this is the area's only function. They prefer a
view that this general area of the brain is an association area that inte-
grates inputs from several centers. They question the evolutionary plau-
sibility of a dedicated region for Word Form analysis. Another challenge
to the VWFA comes from studies using magnetoencephalography

(MEG). MEG is a brain-imaging method with good temporal resolution. In a recent study, Pammer et al. (2004) have used MEG to document the time course of activation in different brain regions during a lexical-decision task. Contrary to the view that the VWFA is a pre-phonological stage of word analysis, they found that activity in the VWFA lagged behind activation in the left inferior frontal gyrus, a region associated with phonological encoding.

An interesting possibility is that the neural networks responsible for written word recognition, including the role of the VWFA, may differ across cultures, depending on the nature of the relationship between the printed symbols and phonology. For instance, Italian has a much more regular orthography than English, that is, Italian has very few exceptions in the correspondence between printed symbols and sounds. Paulesu et al. (2000) compared English-speaking and Italian-speaking college students in tests of word reading and pronounceable non-word reading. The Italian students were faster in their responses, and the patterns of left hemisphere brain activation, measured with positron emission tomography (PET), were different in the two groups. The English speakers showed greater activation in a left posterior inferior temporal region possibly including the VWFA.

We can summarize by saying that the theoretical construct of a visual word-form representation is well established and compatible with bottom-up models of word recognition and reading. Brain-imaging methods, including fMRI, PET and MEG, are focusing on a site in left fusiform cortex for the transformation from representations of letters and words based on visual features to symbolic representations. It remains to be seen whether this brain region is uniquely dedicated to processing text stimuli, and whether it is truly a first stage in visual word decoding in reading.

Bottom-Up and Top-Down Processes in Word Recognition

Our discussion in this chapter has emphasized the role of sensory mechanisms in reading and the importance of bottom-up visual processing. There is no doubt, however, that top-down linguistic factors play a role. In section 2.2, we reviewed evidence for context effects in which people read sentences faster than random lists of words. Such effects show that people are able to take advantage of the semantic or syntactic relationships between words to facilitate reading.

The *word-superiority effect* is a well-known phenomenon that implies that knowledge of words can facilitate letter recognition. This effect refers to the observation that a letter within a word is recognized with higher accuracy than an isolated letter, even when the context provided by the word is not a cue to identification (Reicher, 1969; Wheeler, 1970). For instance, people are better at discriminating "d" from "k" in the context of the words "word" and "work" than in isolation. E. M. Fine (2001) showed that this effect occurs in peripheral vision. She found that the central letter of a trigram in peripheral vision is recognized more accurately when the trigram comprises a word.[18] In a recently completed study in our lab, Ortiz (2002) also evaluated the impact of the linguistic properties of trigrams on letter recognition. He analyzed letter recognition for trigram stimuli on the horizontal midline, extending left and right of fixation. In agreement with E. M. Fine (2001), he found a word-superiority effect in both central and peripheral vision; the central letter of trigrams was recognized better in words than in non-words. He also studied the influence of the frequency of occurrence in English of pairs of letters, termed bigrams. He found a bigram-superiority effect in which letters in frequent bigrams in English, such as "th," were recognized better than letters in infrequent bigrams, such as "qc." Ortiz's findings imply that at least a portion of the word-superiority effect is due to enhanced recognition of common pairs of letters.

In the previous subsection, we discussed a study by Pelli et al. (2003). They found that efficiency for recognizing individual letters was higher than efficiency for recognizing words, providing evidence that letter recognition precedes word recognition. The word-superiority effect may seem to belie this result. In additional analyses related to the "word superiority effect," Pelli et al. showed that word-recognition performance never exceeds a bounding level determined by efficiency for letter recognition.

Pelli, Burns, Farell, & Moore-Page (in press) used methods similar to those of Pelli et al. (2003) to study the role of features in letters. They measured statistical efficiency for letter recognition (computed from contrast thresholds for letter recognition in luminance noise) as a function of the complexity of letters. "Complexity" is a normalized measure of the amount of contour in letters (contour perimeter length squared divided by ink area). Similar to the inverse relationship between efficiency for word recognition and the number of letters in words described above, they found an inverse relationship between efficiency for letter recognition and complexity of the letters—the more complex the letters, the lower the efficiency. Pelli et al. (in press) interpreted their data in the context of a model in which letter recognition is based on the independent detection of a set of simple features. Although the model is not specific about the geometry of these features, the empirical results imply that the features are simple in form (low complexity), consistent with the response properties of size and orientation-selective cells in early visual cortex.

The two Pelli studies—one dealing with recognition of letters prior to words (Pelli et al., 2003), and the other dealing with detection of features prior to letter recognition (Pelli et al., in press)—provide promising conceptual bridges between sensory coding of features and recognition of words and letters. Coupling these conceptual developments with models linking letter recognition to reading, we

---

[18]Visual-span profiles (see section 3.7) are measured with trigrams, and could be influenced by word-superiority effects. Such effects are likely to be small because only about 300 of the 17,576 (1.7%) possible trigrams, composed of the 26 letters of the alphabet, are words.

may be on the verge of a more general understanding of the bottom-up contribution of visual information to reading.

This hierarchical conception of pattern recognition in reading—feature identification first, then letter identity, then word recognition—may break down when the reliability of visual information deteriorates due to non-optimal viewing conditions, poor text legibility (as in the case of peculiar fonts or handwriting), or visual impairment. Under these conditions, readers may need to puzzle their way through the text, taking advantage of lexical inference and context to interpret the words. Given adequate time, people are quite adept at reading words and text with missing letters or typos. "Can _ou rpad thi_ s_ntenee?" The idea that poor quality of visual information about letters can be circumvented by feedback from knowledge of words is embodied in some models of word and letter recognition including the influential Interactive Activation Model of McClelland and Rumelhart (1981). The point here is not to deny that top-down processes may often play a role, but to emphasize the importance of the bottom-up flow of visual information, especially under optimal visual conditions.

Ultimately, visual information about words and letters in reading must merge with non-visual linguistic information processing. This raises the question of how information in the visual pathway concerning letters and words interfaces with linguistic processing in the brain. This is a very large topic extending well beyond the scope of this chapter and our psychophysical studies. To provide some stepping stones from our research on the sensory basis of reading to the interface with language, we include brief reviews of three related topics. Box 2.1 (in chap. 2) discusses the role of the magnocellular pathway in reading and dyslexia. Box 3.2 is concerned with the visual word form area, and Box 3.3 with Dual-Route theory.

**Box 3.3
Dual-Route Theory**

The process of reading aloud involves transformations from written words to spoken words. A prevailing model is dual-route theory (cf. Coltheart, Curtis, Atkins, & Halter, 1993). This model makes the distinction between cognitive representations of words per se, and the semantic meanings usually associated with those word forms. This is similar to the distinction between a spelling dictionary which lists words but no meanings, and a regular dictionary such as Webster's, that also includes meaning. Dual-route theory hypothesizes two pathways from printed words to their verbal (phonological) representations.[1]

---

[1]According to the version of the dual-route model proposed by Coltheart, Curtis, Atkins, and Halter (1993), there are dual pathways from visual input of print to phonology and speech that do not necessarily require semantic processing, and there is a separate pathway from visual input to semantic analysis that does not involve phonological representations (see their Fig. 2).

One of the dual routes from visual words to phonology is the lexical route, in which visual input from letters matches and activates a lexicon (dictionary) of known visual words (sometimes termed an "orthographic lexicon"). Once a match is found, the entry in the orthographic lexicon links to a corresponding entry in a phonological lexicon, a representation of words, based on phonemes, used for speaking. For example, if the visual pathway delivers the string of recognized letters "thyme," the orthographic lexicon determines that it has a matching entry. A connection is then made to a corresponding entry in a phomological lexicon for properly speaking the word aloud. The second nonlexical route involves a set of look-up rules mapping parts of written words (graphemes) into phonemes (termed Grapheme to Phoneme Correspondences or GPC rules.) The GPC rules are derived from natural frequencies in language (e.g., "ch" would map to a phoneme as in "chase" rather than as in "chasm" because of its greater frequency in natural language. When a visual letter string is presented for reading aloud, information is processed in parallel in both pathways. In most cases of frequent familiar words, the lexical route will complete processing first and determine the speech output.[ii]

The lexical route would be effective in reading familiar words when complete letter information is available, even if the words are irregular in the sense of having graphemes that violate the GPC rules (e.g., "yacht" or "colonel"). The nonlexical route would be useful for reading aloud unknown words or proper names when all the letters are identified (how do you pronounce "cablage?"), or perhaps in supporting recognition of words with only partial letter information.

Evidence for the dual-route theory comes from experiments in cognitive psychology (for a review, see Coltheart, Rastle, Perry, Langdon, & Ziegler, 2001). There is also evidence from studies of dyslexia centering on two extreme forms of this disorder. In one form, the reader has trouble reading irregular words such as "yacht", and tends to make regularization errors. Presumably the difficulty is due to a deficit in the lexical route, necessitating the use of the GPC rules in the nonlexical route. These individuals are normal in their ability to read pronounceable nonwords. This pure form is termed surface dyslexia. In the second extreme form, the reader has trouble speaking pronounceable nonwords, but is able to read aloud both regular and irregular words. This disorder, termed phonological dyslexia, is presumably due to a deficit in the nonlexical route,

---

[ii]The dual-route theory was proposed for reading English and probably does not apply to written languages such as Japanese and Chinese in which individual symbols often represent whole words. Coltheart, Rastle, Perry, Langdon, and Ziegler (2001) pointed out that monosyllabic nonwords cannot be written in Chinese or Japanese Kanji script, making the distinctions between lexical and nonlexical routes for reading moot. On the other hand, Italian is a language that uses the same written symbols as English but has a regular orthography. Unlike English, the pronunciation of almost all the words can be inferred from a small number of grapheme-to-phoneme correspondence rules. Such a language may only require a "single route" rather than a dual route from printed words to phonology.

that is, in the pathway that implements the GPC rules.[iii] Supporting this categorization of dyslexics, there is neuropsychological evidence for people with acquired dyslexia with the pure phonological and surface types (reviewed by Coltheart et al, 1993). There is also evidence from an empirical study of children with developmental dyslexia for subgroups with relatively pure forms of phonological and surface dyslexia (Castles & Coltheart, 1993). Castles, Datta, Gayan, and Olson (1999) used a behavioral genetics approach involving a large sample of twins to infer that the phonological subtype has a greater heritable component than the surface subtype.

In section 3.9, we outlined the letters-to-reading model introduced in R20 (2001). This model needed only a simple lexical matching rule to account for RSVP reading speed in central vision. This Exact Matching Rule requires a perfect match between the output of the visual span (a string of identified letters) and entries in an orthographic lexicon. This matching rule corresponds well with the lexical route in the dual-route model.

But in peripheral vision, where the reliability of bottom-up letter information is lower, a model using Exact Lexical Matching underperforms human reading speed, implying that humans do some form of lexical error correction on the fly. It is possible that the nonlexical route of dual-route theory could play a particularly helpful role in this error correction. For instance, when the reader's peripheral visual span delivers a string of letters not directly matching a word in the lexicon (e.g., "infiuence"), correct phonemes associated with parts of the string might provide clues to the identity of the intended word.

This discussion raises the possibility that the lexical route dominates reading in central vision while the nonlexical route plays a relatively greater role in reading with peripheral vision. If this is the case, there would be an interaction between dyslexia subtype and reading performance in central and peripheral vision. Surface dyslexics would be especially disadvantaged in high-speed reading in central vision in which Exact Lexical Matching (or, equivalently, the lexical route) predominates. In peripheral vision where lexical inference is required (perhaps using the nonlexical route), surface dyslexics would be expected to behave more like normal readers in peripheral reading performance. On the other hand, phonological dyslexics may perform relatively well in central vision, but exhibit unusual difficulty in the lexical inference associated with peripheral vision. Given the large number of children who are diagnosed with developmental dyslexia (prevalence estimates range from 5% to 17.5% according to Shaywitz, 1998), we would expect many to eventually experience central-field loss later in life. It is possible that their abilities to use peripheral vision for reading may depend on which of their dual routes to phonology is most impaired.

[iii]There are other classifications of dyslexia based on descriptive rather than model-based criteria, such as the dysphonetic/dyseidetic distinction made by Boder (1973).

3.10 SUMMARY AND CONCLUSIONS

We conclude this chapter with a brief summary of the most important empirical findings and our theoretical interpretations.

Character Size

People with normal vision can achieve their maximum reading speed over a 10-fold range of character sizes from about 0.2° to 2°. The smallest print size yielding maximum reading speed is termed the critical print size (CPS) and is usually close to 0.2°. This is the size of typical newsprint at a common reading distance of 40 cm. Reading speed slows down rapidly as character size decreases below the CPS. Reading speed also slows down gradually for characters larger than about 2°.

Most people with low vision require magnification to read, that is, their CPS is larger than the normal CPS. It is not uncommon for people with low vision to have a CPS of 2° or more. Nevertheless, most people with low vision can achieve functionally useful reading speeds, given optimal viewing conditions, including magnification of character size to exceed their own CPS.

People with central-field loss (scotomas in the central part of the visual field including the fovea) usually read more slowly and require more magnification than people with at least some visual function within the central visual field. This observation is of broad significance to public health because the leading cause of impaired vision in developed countries is AMD, a disease which frequently causes central-field loss.

Contrast

Reading speed in normal vision is nearly unaffected by contrast reduction down to a critical contrast value, typically 10% or less. Below the critical value, speed declines quickly. Speed-versus-contrast curves for different character sizes differ only by a contrast-scaling factor determined by differences in contrast sensitivity for the character sizes in question. These properties of contrast coding in reading can be accounted for by the contrast-response functions of mechanisms in early visual processing known from psychophysical discrimination and masking experiments, and single-cell studies in primary visual cortex.

The interacting effects of character size and contrast on reading can be related to the properties of the CSF for sine-wave gratings.

For people with low vision, the critical contrast for reading is often much higher than the normal value, with the difference predictable from standard measures of contrast sensitivity. For a subset of people with low vision—mostly those with cloudy optics—contrast deficits are a primary factor in explaining reduced reading performance. The contrast-attenuation model, which equates losses in contrast sensitivity with a reduction in stimulus contrast, explains their reading deficits.

For many others with low vision, particularly those with central loss from retinal disease, reading speed is still depressed even when character size and contrast are not limiting factors.

Contrast Polarity

Reading speed for normally sighted participants is usually unaffected by contrast polarity (black-on-white text or white-on-black text).

A substantial number of people with low vision read faster with reversed-contrast (white-on-black) text. These are usually people with abnormal light scatter in the optics of the eyes. These people benefit from computer displays or electronic magnifiers that can render text as bright letters on a dark background.

Spatial Frequency

The important spatial frequencies for reading are to be found within a one- to two-octave band above one cycle per letter (i.e., a band from about 1–3 cycles per letter). Frequencies below one cycle per letter are not important for reading.

Use of a band of spatial frequencies from one to three cycles per letter in reading is consistent with an early stage of processing by a single spatial-frequency channel for letters of a given size. The role of spatial-frequency channels in reading, however, remains uncertain.

Peripheral Vision

Measurements of RSVP reading speed in the lower peripheral field show that the CPS increases with eccentricity at a rate closely paralleling the growth of acuity size for single letters. But even when characters are enlarged in size to exceed the CPS, maximum reading speeds decrease progressively with increasing retinal eccentricity. For instance, in R18 (1998) we found a six-fold reduction in reading speed between central vision and 20° eccentricity. These results are inconsistent with models that equate central and peripheral vision through size scaling.

Visual Span

The visual span is the number of letters, arranged side-by-side as in text, that can be reliably recognized without moving the eyes. We have developed a letter-recognition procedure (the trigram method) to measure visual-span profiles. These profiles are plots of letter-recognition accuracy as a function of the number of letter spaces left and right of the midline.

Three structural features of the visual field play important roles in determining the size of the visual span—decreasing letter acuity in peripheral vision, crowding between adjacent letters, and decreasing accuracy of position signals in peripheral vision.

Even for best viewing conditions, normally sighted people have a rather narrow visual span extending only four or five letter spaces left and right of the fixated letter. The size of the visual span depends on stimulus factors. The visual span shrinks in size at low contrast, for very small or very large letters, and in peripheral vision. Reading speed shows the same dependence on these stimulus variables and is strongly correlated with the size of the visual span.

The close connection between size of the visual span and reading speed has motivated the hypothesis that the size of the visual span is an important determinant of reading speed. We have developed a computational model that implements this hypothesis. The model shows how empirically measured visual-span profiles have a direct impact on reading speed.

It is likely that reduced visual spans play a major role in low-vision reading difficulty. This is especially true for people with central-field loss who must rely on peripheral vision for reading. Studies of reading in normal peripheral vision reveal that reading is slow, probably because of a reduced visual span.

4

Displaying Text

Gordon E. Legge

What do our findings on the psychophysics of reading tell us about displaying text for readers? Chapter 4 focuses on this question. Modern technology provides us with unparalleled flexibility in the design of text displays. Our growing understanding of the visual processing of text should help us to design and display text whose properties are well suited for a reader's vision.

In chapter 3, we discussed character size and contrast because of the importance of these stimulus properties for understanding the visual mechanisms needed for reading. In chapter 4, we deal with several additional properties of text including font, spatial resolution, color and luminance.

Later sections of the chapter deal with the related problems of display size, and methods for navigating through text. The page-navigation process is particularly important for people with low vision who use magnifiers, and also people with normal vision who use small text displays.

The chapter concludes with a summary of guidelines for displaying text in section 4.8. This summary brings together the empirical findings on text legibility reviewed in chapters 3 and 4.

4.1 TEXT LEGIBILITY

Definition

"Legibility" refers to perceptual properties of text that influence readability. Text which is hard to read because of obscure vocabulary, or complex syntax or meaning may be incomprehensible, but still highly legible. Legibility depends on both local and global properties of text. Local properties refer to characteristics of individual letters or pairs of letters such as font, print size, and letter spacing. Global properties refer to layout characteristics such as line length, line spacing, and page format.

Is "legibility" a physical property of text, or a product of visual processing? Legibility is a good example of a psychophysical variable, such as color or brightness, which is dependent on physical stimulus properties, but is fundamentally determined by characteristics of visual processing. Although the physical rendering of text influences the quality of text images on the retina, the ultimate assessment of legibility depends upon the properties of a participant's perceptual representation. Of course, physical distortions of text can result in perceptual representations that do not support good reading; under these conditions, legibility is low.

Tinker's (1963) influential book *Legibility of Print* surveyed the body of research to that time, much of it contributed by Tinker himself and his colleagues, including D. G. Paterson, at the University of Minnesota. Tinker recognized that legibility of print depends critically on perceptual processing. Nevertheless, in the behaviorist tradition of his time, he focused almost exclusively on the empirical effects of text factors, with little attention to the underlying visual mechanisms. In keeping with this approach, he operationalized the definition of legibility in terms of the measurement process: "Legibility of print has been defined to some degree in terms of the technique employed for measuring it" (Tinker, 1963, chap. 2). As discussed in chapter 1 of the present book, a major goal of the research reported in the articles in our series, and reviewed in chapter 3, has been to explore the role of visual perception in reading. Hopefully, this shift of emphasis has helped to awaken interest in visual processing relevant to legibility.

Given that legibility is not an invariant property of rendered text, it is not surprising that legibility can vary with task demands, viewing conditions and the vision status of participants. For example, there is evidence that the relative legibility of two common fonts—Times and Courier—reverse, depending on character size. In R15 (1996), we measured reading speed as a function of character size for these two fonts. Mean reading speeds for 50 normally sighted participants were 5% faster for Times than Courier for characters larger than the critical print size (CPS). For tiny characters smaller than the CPS and near the acuity limit, reading speeds were up to 50% faster for Courier. If reading speed is used as a measure of legibility (see below), this study indicates that for normally sighted participants, Times is more legible than Courier for moderate print sizes while Courier is more legible than Times for tiny print. This example illustrates that character size influences legibility.

The same study (R15, 1996) showed that for a group of 42 low-vision participants, reading speed was generally faster for Courier than Times, both above and below the CPS. This result makes the point that legibility can depend on the participant's vision status.

The results from R15 (1996) also imply that legibility may be different for reading books and reading road signs. Traffic signage often emphasizes reading at a great distance, presumably placing the characters near the acuity limit. Box 4.1 presents a brief discussion of fonts for highway signs.

Box 4.1
Fonts for Highway Signs

The legibility of highway signs is one domain in which an acuity metric seems particularly appropriate. During driving, the distance at which words can be recognized is of principal importance. In two classic reports, Forbes and his colleagues investigated properties influencing the legibility of highway signs (Forbes & Holmes, 1939; Forbes, Moskowitz, & Morgan, 1950). The font used on U.S. highway signs since the mid-twentieth century (Series E-modified) was based on the findings of Forbes et al., and enshrined in the U.S. Federal Highway Administration's Manual of Uniform Traffic Controls and Devices (MUTCD). Forbes et al. used a legibility index as a metric—the number of feet of viewing distance per inch of letter height at which a sign could be read. A legibility index of 50:1 became the standard. For example, if a highway sign has 12-in. high letters and the greatest distance at which the sign can be read is 600 ft, then its legibility index is 600 ft/12 in. = 50:1. Conversion of this physical letter size (12 in.) and viewing distance (600 ft equals 7,200 in.) to a visual angle yields letters subtending 5.7 min-arc (equivalent to 20/23 Snellen size). This size is close to the normal acuity limit, and is a smaller angular size than is usually encountered in books or other printed documents. Recently, the MUTCD has modified its recommended legibility index to 40 ft per inch (equivalent to Snellen size 20/28) because of concern for the growing number of older drivers.

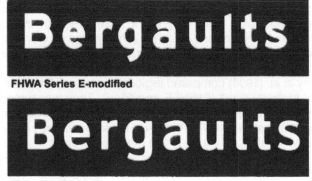

FHWA Series E-modified

ClearviewHwy 5-W

Figure 4.1 Compare examples of the traditional Series E-modified font, used for decades on U.S. highway signs with the new font Clearview Hwy 5-W. Notice that both fonts have ample spacing around the letters to enhance legibility near the acuity limit. Clearview has been designed to accentuate the distinguishing gaps in letters, such as the aperture in e (figure courtesy of Donald Meeker).

The desire to create more legible highway signs has motivated the development of a new typeface for highway signs called Clearview.[1] Clearview is an example of a typeface that has been designed to enhance legibility near the acuity limit.

Two ideas have been paramount in the design of Clearview. The first is to mitigate the effects of a phenomenon variously known as *irradiation, halation,* or *blooming,* associated with positive-contrast signs. Most highway guide signs have white letters on a colored background, typically white on green.

In chapter 3 (section 3.3 and Table 3.2), we referred to white letters on a dark background as "reversed contrast," but in the literature on highway signs, it is termed "positive contrast." Irradiation is an effect in which the thick white strokes of characters in the standard highway font appear to bleed into the surrounding background, obscuring gaps in letters, and making them appear blurry. Irradiation effects are more pronounced in older drivers, possibly because of greater light scatter in their eye's optics. Highly reflective signs, such as those made from retro-reflective material, appear to cause greater irradiation effects. Clearview reduces irradiation by enlarging the interior gaps in letters. This is accomplished by narrowing the stroke width of the font.

The second idea is to emphasize the potential value of lowercase letters for highway signs, and to design the letter forms and interletter spacing to enhance their legibility. In section 3.9, we reviewed the purported advantage of lowercase letters in conveying word shape information. One potential cue to letter shape, the profile of ascenders and descenders, rarely specifies a word uniquely (Walker, 1987). However, it has been pointed out that reading a highway sign often involves distinguishing a word (e.g., street name or city name) from a small set of possible alternatives. This may be a case in which the profile of ascenders and descenders, or some other image property of a word, may provide a unique feature for identification (Garvey, Pietrucha, & Meeker, 1998).

Garvey et al. (1998) compared legibility distances for Clearview and standard highway signs (Series E, modified). They measured legibility distances for 48 subjects from age 65 to 83 years. The subjects rode as passengers in a slowly moving car toward a sign until they could identify the word on the sign, or specify which of three words (located on the top, middle, or bottom of the sign) was a specified target word. The legibility distances did not differ between Clearview and the standard highway font in daytime testing. At night, the recognition distance was about 16% greater for Clearview. Garvey et al. translated this finding into a real-world example. Suppose a standard highway sign is legible at

---

[1]On September 2, 2004, the U.S. Federal Highway Administration issued an interim adoption for the use of Clearview on signs with positive contrast, typically white letters on a green background (Donald Meeker, personal communication, October 4, 2004).

1,000 ft." A 16% increase in legibility distance would correspond to 160 ft. For a car traveling 55 miles per hour, 160 ft is covered in about 2 sec. This means that the 16% increase in legibility distance translates into an extra 2 sec for the driver to read the sign and make a travel decision.

"In this example, if the letters are 24 in. high, the corresponding legibility index for a viewing distance of 1,000 ft is about 42 ft per inch.

Legibility Metric

Tinker (1963, chap. 2) reviewed and evaluated metrics for measuring text legibility. He identified three general types: (a) physiological measures such as blink rate or saccade length; (b) threshold measures, such as acuity size or exposure time; and (c) reading speed. Tinker preferred reading speed, and made the case for it. He concluded, "While speed of performance is not a perfect technique, it appears to be the most nearly satisfactory method we presently have. It has high reliability and apparently good validity" (p. 31). For reasons discussed at length in chapter 2, we have also relied primarily on reading speed as a behavioral measure of reading performance.

Letter acuity, or equivalently the greatest distance at which letters are recognizable, is a common alternative metric to reading speed. Sets of characters that yield better acuity are considered to be more legible. To generalize this metric to legibility for text not at the acuity limit, it is assumed that letters whose critical features can be resolved with the smallest visual angles comprise the most legible text at any scale. There are two reasons to doubt this assumption. First, Chung et al. (2002) and Majaj et al. (2002), discussed in section 3.5, showed that different portions of a letter's spatial-frequency spectrum are used for recognizing letters of different sizes; letter recognition is not scale invariant. If different features are used for recognition for different letter sizes, it is risky to assume that a legibility assessment for tiny print applies equally for all letter sizes. Second, the results cited earlier in this section from R15 (1996) indicate that for normally sighted participants, an acuity-based metric of legibility predicts that Courier is more legible than Times. While this agrees with a reading-speed metric for tiny letters, a reading-speed metric predicts the opposite for letters above the CPS.

In addition to these concerns, the acuity metric has two inherent problems: it cannot be used to study the impact of letter size on legibility, and it is not well-suited to studying the impact of layout properties of text such as word spacing and line length. Reading speed can handle these two problems and has greater face validity than the acuity metric.

Reading speed also has limitations in legibility assessment. These include individual differences in reading skill due to nonvisual factors (education, cognitive status, motivation, and native language), performance differences due to contextual factors, and the challenge of finding a sufficient amount of suitable text for legibility testing. It would be desirable to find a surrogate for reading speed that reduces or eliminates these problems.

4.2 FONT EFFECTS IN NORMAL AND LOW VISION

Influence of Font[1] Design

The effect of font variations on reading speed has been difficult to study for two reasons. First, prior to the advent of computer-based digital fonts, typesetting was done with metallic type or photocomposition. In both of these technologies, it is difficult to manipulate font parameters for experimental purposes.

Second, a particular font design is a creative work, typically not amenable to quantitative stimulus description, and represents aesthetic, functional and other considerations of the designer.[2] The terminology used by type designers is not easily translated into the stimulus dimensions used in vision research.[3] Glossaries of typographical terms can be found in Lawson and Agner (1990) and Bringhurst (2004). Bringhurst also provides a readable discussion of the elements of typographic style.

A difference between two fonts may be specified for a single stimulus dimension, but, pairs of fonts usually differ in many ways, making it difficult to provide a concise quantitative description of the stimulus difference. As an example, the font designer Charles Bigelow (personal communication, June 14, 2005) describes some of the differences between Courier and Times:

> At the same point size, the x-heights are usually different (depending on which versions of Courier and Times are being used). Also, the capital heights are different, ascender heights and descender depths are different, stem and hairline weights are different, the modulation between thin and thick strokes is different, the spacing mode is different (fixed-pitch vs.

---

[1]The term "font" traditionally referred to a particular size and member of a typeface family, such as 12-pt Times Roman. In modern everyday usage, "font" includes different sizes and often bold and italic variants of a typeface family.

[2]The motivation of font designers is an interesting topic in its own right. In a famous essay, Warde (1955, p. 16) emphasized the primary importance of the functional role of type over its aesthetic qualities in conveying ideas from the author to the reader without intrusion upon awareness of the characteristics of the type itself. "The type which, through any arbitrary warping or excess of colour, gets in the way of the mental picture to be conveyed, is a bad type." Many contemporary type designers violate this principle, by creating fonts that are intended to express an attitude above and beyond the verbal message. Although legibility is a priori a central issue in design, it rarely seems to be an explicit goal except for special-purpose fonts such as those designed for roadway signs, or for low vision. Century Schoolbook, designed by Morris Fuller Benton in 1918-19, is an example of a font for which legibility was a major concern. Benton consulted experts at Clark University regarding legibility and children's eyesight to guide his design (Charles Bigelow, personal communication, July 6, 2005).

[3]Typographers and vision researchers sometimes use different terms for roughly the same concept, or the same term for very different concepts. For instance, typographers use the term "weight" to refer to factors such as stroke width that determine boldness or apparent contrast of text, but use "contrast" to refer to the difference between thin and thick strokes of fonts.

proportional), the spatial frequencies are different (Times having a higher but more stable frequency, Courier having a lower frequency overall, but more variable frequency—compare words like "minimum" and "illusion"). By spatial frequency here, I do not mean cycles per letter, but rather cycles per some fixed unit of distance.

Moreover, it is problematic to adopt the standard psychophysical method of parametrically modifying one stimulus property of a font because these modifications might disrupt the underlying design rationale for the font.

For these theoretical and practical reasons, there has been relatively little study of the effects of font properties on reading since the review by Tinker (1963).

The advent of digital typography has facilitated the design and evaluation of new fonts for a variety of purposes, including psychophysical research. Aires Arditi and colleagues at Lighthouse International in New York have created digital fonts for psychophysical legibility assessment. They have used the METAFONT language (Knuth, 1986) to create stimuli in which several font characteristics are varied in a controlled way. They have relied primarily on an acuity metric of legibility, but have also used reading speed. Table 4.1 provides a summary of some of their findings.

Morris, Aquilante, Yager, and Bigelow (2002) did find an effect of serifs on reading speed, unlike Arditi and Cho (2000a, 2005). These investigators used specially designed versions of the Lucida typeface family (Bigelow and Holmes Inc., San Jose, CA), one with serifs and one without serifs. Although typical serif and sans-serif fonts differ in a variety of attributes, the specially designed Lucida fonts were almost identical, apart from the presence or absence of serifs. According to Morris et al. (2002):

TABLE 4.1
Some Font Attributes Studied by Arditi and Colleagues

Attribute	Major Finding	Source
Stroke Width	Optimal range: 0.1 to 0.2 letter height	Arditi et al. (1995)
Aspect Ratio: Height/Width (H/W)	Tall, thin letters (H/W > 1.0) are more legible than short, wide letters (H/W < 1.0)	Arditi (1996)
Serifs: Presence or Absence	No effect on reading speed, and tiny improvement of acuity with serifs	Arditi & Cho (2000a, 2005)
Solid and Outline Letters	Acuity size for outline letters is about 1.82 times larger than for corresponding solid letters.	Arditi et al. (1997)
Fixed Width (FW) vs. Proportional Width (PW)	For letter size near the acuity limit, reading speed is faster for FW. For larger print, reading speed is faster for PW	Arditi et al. (1990)

The designers produced a seriffed and sans-serif pair whose underlying forms are identical in stem weights, character widths, character spacing and fitting, and modulation of thick to thin. The only difference is the presence or absence of serifs, and the slight increase of black area in the seriffed variant. (p. 245)

The investigators used the RSVP method to measure reading speed for the two font variants. For a small print size of 12 min-arc (probably near the CPS), reading speed was about 20% faster for the sans-serif font, but there was no difference in reading speed at a larger print size of 48 min-arc.

The wide range of typeface designs, alphabets, and nonalphabetic writing systems in use worldwide indicate that there is wide latitude in the details of reading symbols that are legible for human vision. Some recent psychophysical evidence suggests that the elementary sensory features used by reading vision are spatially simple and probably common across different writing systems. As discussed in section 3.9, Pelli et al. (2003) measured the statistical efficiency with which normally sighted participants recognize low-contrast letters imbedded in visual noise compared to an ideal letter-recognizing algorithm (termed "ideal observer"). They measured recognition efficiency for several English fonts (Bookman, Courier, Helvetica, Kunstler & Sloan), several non-English scripts (Arabic, Armenian, Chinese, Devanagari, & Hebrew), and also artificial alphabets made up of checkerboard patterns including one similar to Braille. In all cases, the participant's task was to determine which character was presented from a set of 26 alternatives. They found that recognition efficiency was about 10% for traditional alphabetic fonts, lower for more complicated symbols sets (Kunstler font, Arabic, & Chinese), and higher (30%) for a simple 2×3 checkerboard font similar to Braille. Remarkably, a single quantitative measure of pattern complexity (squared contour length divided by ink area) accounted for the variation in efficiency across these symbol sets; efficiency was inversely proportional to pattern complexity. By extrapolation from their data,[4] they estimated that only very simple pattern features, such as disks or short segments of contours, are recognized with high efficiency approaching 100%. They proposed a model in which letter recognition depends on the independent detection of several such simple sensory features; more complex alphabets or fonts are recognized with lower efficiency because more elementary features must be detected.

---

[4]From the definition of pattern complexity used by Pelli et al. (in press), the simplest (minimum complexity) pattern is a disk, for which ratio of squared contour length to area is $(2\pi r)^2/\pi r^2 = 4\pi \approx 12.6$. By the same calculation, a square has a complexity of 16. Pelli et al. found that an artificial checkerboard font, consisting of arrangements of squares in a 2×3 layout yields an efficiency of about 30%. Assuming that the symbols in this checkerboard font have a complexity equivalent to about three nonaligned squares (complexity = 48), and that the inverse relationship between efficiency and complexity holds, we would predict that recognition efficiency is close to 100 percent for a simple pattern consisting of one square. Following similar reasoning, Pelli et al. (in press) inferred from their data that human vision is specialized for detecting very simple pattern features as the building blocks of letter recognition.

Although the model proposed by Pelli et al. (in press) is not specific about the geometry of the elementary features, their data imply that the features could be simple blobs, or oriented lines or edges well-matched to the receptive-field centers of cells in primary visual cortex. If this analysis is right, then visual letter recognition does not rely on complex letter features, but configurations of very simple features. A consequence is that the visual requirements for font design would be very loose; any set of characters should do, as long as they contain a sufficient number of simple, detectable features in distinct spatial configurations.

Influence of Spacing

Arditi, Knoblauch, and Grunwald (1990) compared reading speeds for a proportionally spaced font (Times Roman) and a fixed-width variant created by adding empty space around the narrower letters to equalize the horizontal real estate per letter. They found that reading speed was faster for the fixed-width font for tiny letters near the acuity limit. They attributed the slower reading of the proportionally spaced font to greater crowding between letters. For medium and large character sizes, reading was faster for the proportionally spaced font. This difference might be due to requirements of eye-movement control. Most of their experiments involved reading single lines of text on a computer screen, a task which required saccadic eye movements. When they used RSVP presentation, in which the need for eye movements is reduced, the reading-speed difference between the fixed-width and proportionally spaced fonts disappeared for the medium and large print sizes.

We also became interested in the influence of fixed-width and proportional spacing on reading, in connection with the design of the MNREAD acuity chart (see chap. 5). The original version of this chart used Courier, a fixed-width font. Subsequently, the MNREAD acuity chart was redesigned using a proportional font, Times Roman, because most modern printed material uses proportional fonts. Our interest in the Courier and Times versions of the MNREAD chart, and the findings by Arditi et al. (1990) on the different effects of fixed and proportional spacing on reading speed, motivated us to compare Times and Courier reading performance in both normal and low-vision participants (R15, 1996). We found that maximum reading speeds differed only slightly for the two fonts. For 50 normally sighted participants, the average maximum reading speed was 5% greater for the proportional font. This small difference (compared with the individual variability in reading speeds estimated in Table 2.1) is consistent with font variations in reading speed measured by Arditi et al. (1990) and Klitz, Mansfield, and Legge (1995). For the 42 low-vision participants, many with macular degeneration, maximum reading speeds averaged 10% faster for the fixed-width font. For both normal and low-vision participants, reading speeds for small letters near the acuity limit were substantially faster for the fixed-width font.[5]

---

[5]Reading acuity and critical print size (defined in section 3.2) both corresponded to smaller letters for Courier than Times in both normal and low vision.

Our general interpretation of the foregoing font results, consistent with the conclusions of Tinker (1963, chap. 4), is that type designers have developed several commonly used fonts that are roughly comparable in terms of reading performance for normal vision, at least when angular character size is greater than some critical value. For low vision, fixed-width fonts may yield faster reading, possibly because low-vision reading often occurs near the acuity limit.

The advantage of fixed-width fonts near the acuity limit suggests that fonts with extra-wide letter spacing might be advantageous for reading, especially for low vision. The extra space might relieve crowding effects between adjacent letters, likely to be especially severe in peripheral vision. Contrary to this intuition, there is very little evidence that extra-spaced fonts yield fast reading. In R2 (1985), we reported on a limited experiment in which we measured reading speed (drifting-text method) for two normal and four low-vision participants. They read highly magnified text (character size of 6° or larger), with normal spacing, and 1.5× and 2× normal spacing. For all of the participants, reading speed was highest for normal spacing and declined for extra spacing (R2, 1985, Fig. 8).

Chung (2002) used the RSVP method to make extensive measurements of reading speed for six normally sighted participants. They were tested foveally and at 5° and 10° in the lower visual field with two print sizes (above and below the CPS). The Courier letters had horizontal spacing ranging from 0.5× to 2× normal spacing. In all conditions, reading speed was maximum at or near the normal spacing and was never greater for extra-wide spacing than for normal spacing. Chung also tested five low-vision participants with AMD and found no advantage in reading speed for extra-wide spacing (Chung, 2005).

Berger, Martelli, Su, Aguayo, and Pelli (2003) have reported that extra-wide spacing of letters yields faster reading speeds in peripheral vision for strings of random words, but not for sentences. They conclude that extra-wide spacing can reduce the impact of crowding in peripheral reading, but that top-down effects of word order overwhelm this effect in continuous text.

The bottom line is that the more ample spacing around narrow letters in a fixed width font compared with a proportionally spaced font appears to facilitate reading in low vision and also near the acuity limit for normally sighted people. But adding extra spacing to standard fonts does not have practical advantages for reading in normal or low vision.

Fonts for Low Vision

Digital font design has made it realistic to consider special-purpose fonts for low-vision readers. The Tiresias[6] family of fonts, developed at the Royal National Institute of the Blind in London, England, includes fonts intended to optimize legibility for different applications, including television subtitles, computer displays

---

[6]Tiresias was a blind prophet in Greek mythology.

and cash dispensing machines, public signs, and for large print books (Gill & Perera, 2003). The design of the large-print version of Tiresias was guided by a series of surveys of potential users soliciting their preferences for font characteristics (Perera, 2001). Similarly, the American Printing House for the Blind (APH) has designed a font for low vision called APHont (Kitchel, 2004). In keeping with the research reviewed above, both Tiresias and APHont emphasize ample space around narrow letters.

It may not be possible to find a single best low-vision font, given the heterogeneity of eye conditions. Arditi (2004) has explored the intriguing possibility of individually customizing fonts. In Arditi's (2004) study, 40 low-vision participants were tested with a program called Font Tailor. The program enabled independent adjustment of several font parameters—letter spacing, stroke width, serif size, x-height (relative to the height of capital letters) and aspect ratio. The participants viewed samples of text and separately adjusted the font parameters to optimize their judgment of legibility. Once the participant's parameters were chosen, legibility was measured objectively using the acuity metric. Arditi (2004) found substantial individual variations, confirming the view that customized font design may be helpful in low-vision reading. Across participants, font adjustment resulted in substantial improvement in legibility, although the resulting individualized fonts did not exceed the legibility of Times Roman. While encouraging, this research leaves open two questions: Can individual tailoring produce fonts yielding better low-vision reading performance than standard fonts such as Times, Arial or Courier? Are there principled ways to guide the tailoring of fonts for people with different forms of low vision?

4.3 DISPLAY RESOLUTION

How does the amount of detail with which letters are rendered affect the legibility of text? As recounted briefly in the Preface, this is the question that first stimulated our interest in visual factors in reading. Our original interest in this issue was motivated by design requirements for low-vision reading displays (R1, 1985; R2, 1985; R3, 1985). Subsequently, this question has emerged as a central topic in the design of text displays for normal vision as well. Since the 1980s, there has been widespread adoption of technology for displaying text as a set of samples, including pixel-based graphics, matrix printers, and bitmap printers (e.g., laser and inkjet printers.) For all of these technologies, the sample density (dots or pixels per unit distance) is a primary specification because it determines the resolution (samples per letter) for rendering letters of different sizes.

Effects of Sample Density and Spatial-Frequency Bandwidth

We evaluated the effect of spatial resolution in two ways, by measuring reading speed for sampled text, and spatial-frequency filtered or "blurred" text (R1, 1985). Figure 4.2 shows examples of sampled text and Figure 4.3 shows examples of blurred text.

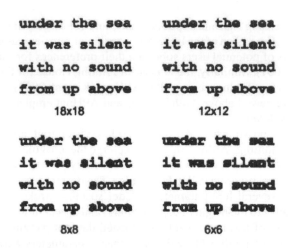

Figure 4.2 Examples of sampled text. The four panels show sentences rendered with sample densities of 18 × 18, 12 × 12, 8 × 8, and 6 × 6 samples per character (samples per x-height).

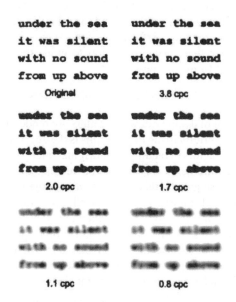

Figure 4.3 Examples of blurred text. An original sentence is shown, and five low-pass spatial-frequency filtered versions. The filtered versions were created in Matlab with Gaussian low-pass filters with bandwidths of 3.8, 2.0, 1.7, 1.1, and 0.8 cycles per character. Empirical findings in R1 (1985) indicate that a bandwidth of 2 cycles per character is all that is required for maximum reading speed.

In the late 1970s, prior to the availability of computer-addressable video frame buffers and image-processing software, we used optical techniques for sampling and filtering. Sampled displays were created by placing an opaque, black acetate sheet with a regular array of transparent holes over the TV screen. Different acetate sheets contained arrays of different sample densities, ranging from 1.4×1.4 to 22×22 samples per character. Low-pass spatial-frequency filtering (blur) was produced by interposing a diffuser—a large sheet of ground glass—between the observer and the TV screen. The filter bandwidth[7] (amount of blur was determined by the distance of the diffuser from the screen and was adjusted by means of a manual, screw-driven positioner.

We measured reading speed as a function of sample density and spatial-frequency bandwidth using the drifting-text method. As illustrated in Figure 4.4 (reprinted from R1, 1985, Fig. 5), we found critical sample densities and blur bandwidths for reading—below the critical values, reading speed slowed down, and above the critical values, reading speed was independent of sample density or spatial-frequency bandwidth.

Figure 4.4 shows that the critical sample density (CSD) for $1.5°$ letters is about 8×8 samples per character and the critical blur bandwidth is about 2 cycles per letter. We return to the effect of character size on these critical values below.

How does the CSD of 8×8 compare with pixel resolutions of computer displays? Table 4.2 shows sampling resolution for 12-pt Times New Roman letters, rendered with two display configurations—an 800×600 pixel display with 60 pixels per inch, and a $1,280 \times 1,024$ pixel display with 90 pixels per inch. For both displays, the uppercase and lowercase letters have pixel heights of 9 and 6, close to the CSD of 8×8 just discussed.[8]

---

[7]The spatial-frequency bandwidth of the filter is specified in units of cycles per letter. For example, a bandwidth of 4 cycles per letter means that the contrast of sinewave gratings with spatial frequencies above 4 cycles per letter would be severely attenuated by the filter. Characterization of the filtering properties of the ground-glass diffuser was accomplished optically as described in the Methods section of R1 (1985). The filter bandwidth in cycles per letter can be converted to units of cycles per degree if the angular size of the letters is known. For instance, when the letters subtend $0.5°$, a filter bandwidth of 4 cycles per letter is equivalent to a filter bandwidth of 8 cycles per degree. Expressing the bandwidth in cycles per letter (sometimes termed "object" spatial frequency) is relevant to describing how much spatial detail about the letter is displayed, while expressing the bandwidth in cycles per degree (often termed "retinal spatial frequency") allows the observer's contrast sensitivity function to be taken into account.

[8]For the examples in Table 4.2, the 12-pt letters retain the same height in pixels for the two screen resolutions, but the physical (and angular) size of the letters decreases for the higher-resolution display. This peculiar behavior seems to defy the standard notion of point size as an objective, physical measure of print size (see Appendix A). What is going on?

It is likely that the Windows operating system uses the PostScript definition of point size (see Appendix A)—that is, 72 points per inch. The question is how the operating system takes into account properties of the display—its underlying type (CRT, LCD, plasma, etc), the physical dimensions of the image, and the number of pixels in the display—in rendering letters from a given font at a given point size. (The application itself may add another variable if it permits zooming of print size.) For the conditions of the

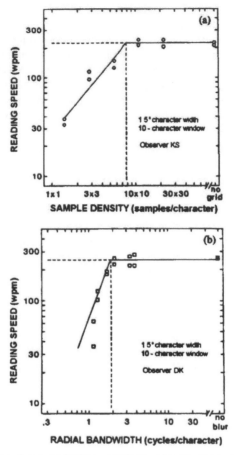

Figure 4.4   Effects of sample density and blur bandwidth on reading speed. (a) Reading speed is plotted as a function of sample density for observer K.S. The data have been fit with a rising straight line and a horizontal straight line. The point of intersection of the two lines has an X coordinate of about 8 × 8 samples per character and is called the critical sample density (CSD). The corresponding Y coordinate is about 220 wpm and is the maximum reading speed. (b) Reading speed is plotted as a function of radial, spatial frequency bandwidth of a ground-glass diffuser for observer D.K. The maximum reading speed of 247 wpm is reached at a critical bandwidth of 2.0 cycles per character (From R1, 1985, Fig. 5).

---

examples in Table 4.2, the operating system is apparently treating pixel count as a more stable indicator of print size than other display characteristics, so the upper- and lowercase x-heights remain constant in pixels for the two screen resolutions. By comparison, printers usually have fewer settings to take into account, so the size of print in hardcopy rendering is more predictable.

Even if the display rendering is accurate, it is still hard to predict the relationship between x-height and point size for a font. A rule of thumb is that for most well-behaved designs, the lowercase x-height in a 12-pt font is between 4-pt and 7-pt (Charles Bigelow, personal communication, Sept. 12, 2004).

To reiterate the recommendation in Appendix A, if character size is a relevant variable in research involving computer-generated text, it is advisable to directly measure x-height for the display settings being used.

For the reading distance of 16 inches (40 cm) in Table 4.2, the upper and lower-case x-heights subtend 0.54° and 0.36° for the 800 × 600 pixel display (smaller for the 1,280 × 1,024 display). This is well within the optimal range for reading, but smaller than the 1.5° size for which the CSD of 8 × 8 was measured. Does CSD or critical bandwidth depend on character size?

Our measurements showed that the critical bandwidth for reading is constant at 2 cycles per letter, independent of character size (R1, 1985, Fig. 7b). This apparently low bandwidth corresponds to visibly blurry text (see Fig. 4.3). This result means that blurry, low-resolution letters are sufficient for rapid reading. Recalling our discussion of font characteristics in the previous section, Blur of this severity will reduce or eliminate the visibility of some of the fine details that distinguish one font from another such as the shape of the serifs or the variation of stroke widths. So, while a higher spatial-frequency bandwidth may be necessary to distinguish between fonts or appreciate their fine details, these high-bandwidth features appear to be unnecessary for rapid reading.

From the sampling theorem, information in a character that is band-limited to N cycles per character can be represented by 2N × 2N samples. Given the character-size invariance of critical bandwidth at 2 cycles per letter, sampling theory predicts that critical sample densities would be constant at 4 × 4 samples per letter. Our measurements revealed critical sample densities of 4 × 4 for tiny letters near the acuity limit. Surprisingly, however, the CSD gradually increased, from about 4 × 4 for the tiniest letters, up to about 20 × 20 samples for letters larger than 10° See the solid line in Figure 4.5, reprinted from R1 (1985, Fig. 7A).

These findings on the character-size dependence of the CSD were obtained with the drifting-text method and matrix sampling. Modern electronic text displays typically use static text and pixel sampling. In a recent experiment in our laboratory by Deyue Yu, we have conducted a partial replication of the original experiment but with static text and pixel sampling. Reading speeds were measured for three young, normally sighted participants using the flashcard method

TABLE 4.2

Example of Sampling Resolution for Letters on a Pixel Display*
(Font: Times New Roman, 12 pt, viewed from 16 inches)

Screen Dimensions		Screen Resolution (pixels per inch)	Upper Case X-Height		Lower Case X-Height	
Inches	Pixels	Pixels	(°)	Pixels	(°)	Pixels
13.3 × 10	800 × 600	60	0.54	9	0.36	6
14.2 × 11.4	1280 × 1024	90	0.36	9	0.24	6

*Values in this table were obtained as follows: A CRT display was adjusted so that the screen resolution was 60 pixels per inch for a display of 800 × 600 pixels. The operating system was Windows 2000. Matlab Version 6.5 was used for displaying letters. After measuring pixel and angular heights for this condition, the Windows Control Panel was used to change the display to 1280 × 1024 pixels for the second set of measurements.

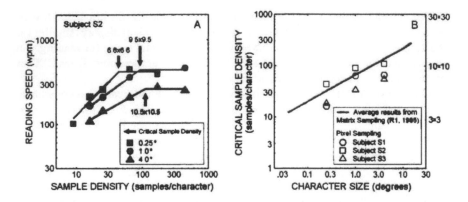

Figure 4.5 Results of an experiment with pixel sampling and static text. Reading speed was measured as a function of sample density. The flashcard method was used to measure reading speed (section 2.1). A: Data for one participant, showing reading speed versus sample density for three character sizes. Arrows indicate the CSD for each of the three character sizes, expressed in linear units. For instance, the CSD at 0.25° is 6.6 × 6.6 pixels, equivalent to 43.6 samples/character. B: The data points show CSD as a function of character size for three participants in an experiment with pixel sampling and static text. The solid line replots average values of the CSD from R1, 1985, Figure 7A, obtained in experiments with matrix sampling. *Note.* In this figure, "character size" is defined as x-height. In R1 (1985), We used the center-to-center distance between characters as our metric of character size and used the term "character width." For the Courier font used in the pixel sampling experiment reported in the present figure, character width is 1.57 times larger than x-height. This conversion has been used to plot the matrix-sampling results (solid line) in panel B.

(section 2.1). Each participant was tested at three character sizes—0.25°, 1.0°, and 4.0°— and five sample densities for each character size. The samples were square black and white pixels. Examples of the stimuli are shown in Figure 4.2. For each character size and sampling condition, reading speed was determined from a psychometric function of percentage of words read correctly as a function of sentence exposure time (there were five test sentences at each of five exposure times, and reading speed was computed from the exposure time yielding 80% correct.)

Figure 4.5A shows plots of reading speed versus sample density at the three character sizes for participant S2. Consistent with the sampling results in R1, the critical sample densities for participant S2 increase as character size increases. The data points in Figure 4.5B show critical sample densities as a function of character size for three participants in our pixel-sampling experiment. The solid curve replots average results from R1 (1985, Fig. 7A) for matrix sampling. The reasonable agreement between the two data sets confirms that the character-size effects observed in R1 (1985) generalize to the reading of static text and pixel sampling.

In R1 (1985), we considered explanations for the unexpected growth in CSD for large characters. We favored a masking account, similar to one proposed by Harmon and Julesz (1973) to explain the block portrait effect. In the case of reading, the idea is that high spatial-frequency components produced by the edges of coarse samples would mask information in the important spatial-frequency band for reading (information up to 2 cycles per letter). For tiny letters, these masking artifacts would be unimportant because they would be attenuated by the high-frequency decline in contrast sensitivity. Only for tiny letters would the low CSD predicted from sampling theory apply. For large letters, these artifacts could be pushed to higher frequencies, remote from the band for reading, by increasing the sample density. To accomplish this, the CSD for large letters would be higher than predicted from sampling theory.

Majaj et al. (2002) proposed an alternative explanation that does not rely on masking. As discussed in connection with the spatial-frequency channel model of reading (section 3.5), these authors found that the critical spatial-frequency information for letter recognition resides at higher frequencies (in units of cycles per letter) for large letters compared with small letters. For large letters, recognition relies on high-frequency spectral components associated with edges. These components require a high sampling density to represent them. For tiny letters, near the acuity limit, vision relies on gross blobs, rather than edges, requiring much lower sampling density.[9] This account plausibly explains why CSD depends on character size, but still leaves open the question of why the critical blur bandwidths do not show the same dependence on character size.

Implications for Text Displays

There are two implications of our finding that CSD, but not critical bandwidth, depends on angular character size. First, angular character size depends on the reader's viewing distance which is often unconstrained. To allow for the widest range of viewing distances, the sample density should be at least 20 × 20 samples per character.

Second, blurring, digital or optical, may enhance the legibility of coarsely-sampled text displays. Anti-aliasing (or grayscale smoothing) is a closely related method for improving the appearance of text characters rendered on pixel arrays such as cathode ray tube (CRT) or liquid crystal displays (LCDs). When text is rendered using only light and dark pixels (termed "bilevel" or binary displays), character features such as diagonal strokes or curves can be distorted. The stair-stepping effect on the contour boundaries (see the examples in Fig. 4.2), are referred to as "jaggies." Anti-aliasing takes advantage of the ability of many displays to render pixels with more than two gray levels. For a brief review of anti-aliasing

---

[9] In their paper, Majaj et al. (2002) demonstrated quantitative agreement between the dependence of critical sample density on character size from R1 (1985) and their measurements of optimal spatial-frequency bands as a function of letter size (see their Fig. 6). To make this comparison, they took half of the critical sample densities from R1 as estimates of the optimal channel frequency.

methods in relation to human vision, see Morris, Hersch, and Coimbra (1998). The traditional method of anti-aliasing is to (a) low-pass spatial-frequency filter an original high-resolution bilevel representation of the text, which converts the original bilevel image into a blurry grayscale image, and then; (b) resample this blurry image for display on the pixel array. Gould, Alfaro, Finn, Haupt, and Minuto (1987, Experiment 5) found no significant advantage in reading speed for this type of anti-aliased text over bilevel text rendered with the same pixel resolution. A more sophisticated approach to anti-aliasing, termed "perceptual tuning" or "hinting," focuses on preserving the integrity of important structural features of letters (e.g., enhancing thin contours) in converting from bi-level to grayscale. The adjustments can be implemented by algorithm or by manual fine tuning of a font. O'Regan, Bismuth, Hersch, and Pappas (1996) found no advantage for perceptually tuned grayscale fonts over bilevel fonts for reading speed, but there was a small advantage in a character-string search task. These results indicate that anti-aliasing has little or no effect on reading speed.[10]

Why does anti-aliasing not improve reading speed for text with jaggies? The answer probably relates to angular character size. The studies have focused on small characters on computer displays (6, 8 or 10 pt) because these are the characters that are coarsely sampled. But at a normal reading distance, these characters subtend a small visual angle.[11] Our findings in R1 (1985) indicate that the critical sample densities become quite small for tiny characters. If the sample density of a bilevel font exceeds the CSD, then anti-aliasing would be unlikely to result in faster reading. This analysis implies that anti-aliasing would be more valuable for large, coarsely sampled text.

Several early studies indicated that reading from CRT displays is slower than reading from paper, perhaps due to intrinsic display properties such as raster flicker. For citations, see Gould et al. (1987), and Jorna and Snyder (1991). More likely, however, the difference was due to sampling differences between CRT text and printed text. These authors showed that reading speeds are equivalent for CRT and hard-copy when spatial resolution is adequate and characteristics of letters are match such as size and font (Gould et al., 1987), or when the spatial-frequency content of text images are matched (Jorna & Snyder, 1991).

Sub-pixel addressing is a method for extending the effective sampling resolution for those displays for which a given display pixel is actually composed of horizontally adjacent red, green and blue "sub-pixels." This is the case for many LCDs. If the luminance of these sub-pixels can be controlled independ-

---

[10]An exception to this conclusion was reported by Sheedy and McCarthy (1994). In their study, text images were acquired by a scanner at one resolution and displayed on a monitor at a lower resolution. The displayed images were either bi-level (black and white pixels only) or contained gray levels. Reading speeds were faster for the gray-level displays. This study differs from the other cited anti-aliasing studies in using scanned images as the starting point from which bi-level and gray-level stimuli were derived.

[11]In Gould et al. (1987, Exp. 5), the character height was 3.5 mm. If the subjects viewed the text from 40 cm, the corresponding angular height was 0.5°.

ently, then the effective sampling density in the horizontal direction is increased by a factor of three. The downside of this enhanced luminance sampling is the introduction of uncontrolled chromatic variations (artifacts) at high spatial frequencies. Because human vision can resolve higher spatial frequencies in luminance contrast than in color contrast, the enhanced luminance sampling may be helpful while the chromatic artifacts are invisible. Daly (2001) has analyzed the luminance and chromatic signals associated with sub-pixel addressing in relation to human luminance and chromatic contrast sensitivity data. He has pointed out that sub-pixel addressing is plausibly helpful only for an optimal range of viewing distances. When the viewing distance is too large the extra bandwidth associated with subpixel addressing is not helpful because it lies outside the high spatial-frequency cutoff of the luminance CSF; when the viewing distance is too short, the chromatic artifacts from sub-pixel addressing become visible because they lie below the high-frequency cutoff of the chromatic CSF.

Sub-pixel addressing increases the horizontal sampling density of a display, but not the vertical sampling density. Our study (R1, 1985) always used equal sampling in the horizontal and vertical directions. We did not determine whether extra samples in the horizontal direction could compensate for inadequate sampling in the vertical direction. For instance, if the CSD for 1.5° letters is 8 × 8 pixels, it is not known if an asymmetric arrangement with fewer samples in the vertical dimension but more in the horizontal would be adequate for rapid reading; for example, 6 (vertical) × 18 (horizontal).

Sub-pixel addressing has been implemented by Microsoft Corp. in its ClearType fonts. One study has shown that participants prefer text rendered with ClearType to otherwise identical text rendered on the same display with traditional pixel sampling (Tyrrell, Pasquale, Aten, & Francis, 2001). But the same study found no difference in reading speeds for the two types of pixel sampling.

Defocused Text

We turn briefly to the effect of defocus blur on reading. We distinguish between blur produced in two ways. The first, to be termed "stimulus blur," is to attenuate the high spatial frequencies of the text (low-pass filtering) before presenting it to a well-focused participant. Stimulus blur can be accomplished with digital image processing or with an optical diffuser. We used a ground glass diffuser to produce stimulus blur in R1 (1985) to study the effect of spatial-frequency bandwidth on reading speed. A second type of text blur is produced when a regular text stimulus is presented to a participant who is defocused, either because of intrinsic refractive error or from a defocusing lens in front of the eye. Almost everyone reads defocused text at one time or another; text appears blurry when the viewing distance lies outside the participant's range of accommodation. For a myopic (short-sighted) reader without glasses or contact lenses, distant text (e.g., street signs) will be out of focus. For a hypermetropic (far-sighted) reader, text at a mormal reading distance is likely to be out of focus.

Both types of blur attenuate the high-frequency content of images more than the low-frequency content, a transformation of the retinal image that presumably underlies the perception of blur.[12] Stimulus blur can be described as a transformation of the stimulus, without reference to the participant, and can often be characterized by a linear filter function. Defocus blur depends on characteristics of the participant's visual optics, including pupil size, refractive and accommodative state, and ocular aberrations. In principle, the spatial-frequency filtering effects of defocus can be represented as a component of a linear model of the optics of the eye (cf. Legge, Mullen, Woo, & Campbell, 1987), but in practice, it is rarely possible to know exactly how a given amount of optical defocus modifies the content of retinal images in an individual eye. Empirical studies have shown that the size of acuity letters increase roughly in proportion to the number of diopters of blur (Legge et al., 1987; Thorn & Schwartz, 1990). Four diopters of positive lens defocus can reduce Snellen acuity to about 20/200, and 12 diopters of defocus can reduce acuity to less than 20/1000.[13]

A pair of studies on the recognition of traffic signs provides a hint that defocus blur may have surprising effects on reading. Kline, Buck, Sell, Bolan, and Dewar (1999) used plus lenses to defocus normally sighted young and old participants to Snellen acuity levels of 20/30 and 20/40. Under these defocused conditions, they measured the smallest angular sizes at which their participants could identify text signs, either familiar traffic signs or novel signs of a similar design. Surprisingly, the older participants did better than the younger participants; they could read both kinds of defocused signs at significantly smaller angular sizes. The greater depth of focus of the aging eye, due to smaller pupils in old age, is unlikely to be the explanation because both the younger and older groups were defocused to a criterion level of acuity before legibility testing. One possible explanation for the older participants' advantage is greater experience with defocused stimuli in general. Alternatively, it is possible that ocular changes in aging eyes might differentially influence the properties of defocused images. In a follow-up study, Bartels and Kline (2002) tried to distinguish between these alternatives by testing young and old participants with stimulus blur (produced by digital filtering) rather than defocus blur. This time the younger participants performed better in recognizing blurry words on traffic signs. The authors concluded that the better performance of

---

[12]M. A. Webster, Georgeson and S. M. Webster (2002) have shown that people adapt quite rapidly to blur in such a way that stimuli with abnormally steep drop-offs in their spatial-frequency spectra appear less blurry. This adaptation may underlie the perceptual experience of many people with low vision. Despite low acuity, sometimes due to chronic optical blur or light scatter, the world does not appear subjectively blurry to them.

[13]Legge, Mullen, Woo and Campbell (1987) fit their measurements of letter acuity as a function of positive lens defocus (for subjects with dilated pupils) with the equation, $A = 0.64/D^{1.45}$, where A is decimal acuity and D is defocus in diopters. This equation is equivalent to $MAR = 1.56D^{1.45}$, where MAR is the minimum angle of resolution in min-arc for the critical features of acuity letters.

their older participants with defocus blur was probably due to "compensatory" changes in the optics of the aging eye. The nature of these age-related optical changes remains unknown.

In a recent study, Jarvis and Chung (2004) measured reading speed for defocus blur, produced by lenses, ranging from 0 to 3.0 diopters (D). If a given amount of defocus blur acts like a low-pass spatial-frequency filter, we would expect that by making characters large enough, the critical band of roughly two cycles per character (R1, 1985) should fall within the band of frequencies passed by the filter. In other words, we would predict that for any fixed level of defocus blur, it should be possible to increase character size and eventually achieve maximum reading speed. Surprisingly, this is not what Jarvis and Chung found. For each level of defocus, they measured reading speed as a function of print size for their 19 participants. As defocus increased from 0 to 3.0 D, mean values of the maximum reading speed declined from 170 to 106 wpm.

These studies imply that defocus blur may have different effects on reading than stimulus blur. If so, we need to discover what difference in the two types of blur is responsible. A possible candidate is *spurious resolution*.[14] This is a peculiar effect of a defocused optical system in which the polarity of a contour can reverse. Spurious resolution has been demonstrated to occur in the defocused human eye (Legge et al., 1987). Spurious resolution has been proposed as a contributor to the effects of defocus on letter acuity (Thorn & Schwartz, 1990, but this idea has been challenged by Akutsu, Bedell, & Patel, 2000). Although polarity switches of entire letters do not seem to affect reading speed (see section 3.9), reversal of polarity of contours within letters might have an impact on reading.

Reading and Myopia

How accurately do people focus during reading? Schaeffel, Weiss, and Seidel (1999) used a specially designed autorefractor to measure the plane of focus during reading for twelve young adults. Six of the participants were myopes who wore their proper lens corrections, and six were emmetropes who required no lens correction. For text at a near reading distance of 30 cm, both groups of participants under-accommodated; they focused slightly beyond the text distance with a defocus error of about 0.3 D.

According to the classic psychophysical measurements of the depth of field of the human eye by Campbell (1957), people are near threshold for detecting defocus blur when small, high-contrast stimuli are 0.3 D away from the plane of fo-

---

[14]Spurious resolution occurs when the modulation transfer function (MTF) of an optical system passes through zero and becomes negative. The optical consequence is a phase reversal in the image of a sine-wave grating. Spurious resolution can occur in severely defocused optical systems. An example is shown in Legge, Mullen, Woo and Campbell (1987, Fig. 1). These authors also provided evidence that spurious resolution occurs in the human eye.

cus of the eye.[15] The findings from the Campbell and Schaeffel et al. (1999) studies are consistent with the idea that when people read text at a typical near viewing distance, the eyes accommodate slightly behind the text at a point where defocus blur is no longer detectable visually, remaining slightly out of focus by about 0.3 D.

What effect might such a small amount of defocus blur have on reading? We would expect no effect on reading speed for letters larger than the CPS, based on the findings discussed above on low-pass spatial-frequency filtering and the defocus study by Jarvis and Chung (2004). Although small defocus errors would probably have no impact on reading performance, Wallman and Winawer (2004) have proposed the intriguing hypothesis that defocus in reading may contribute to myopia (see also, Wallman, Gottlieb, Rajaram, & Fugate, 1987).

The putative link between defocus in reading and myopia is based on a developmental process known as emmetropization. A great deal of evidence, reviewed by Wallman and Winawer (2004), has shown that defocus and other types of pattern deprivation trigger growth of the eye in young animals including cats, monkeys and humans. For instance, experiments have shown that when infant monkeys experience long-term defocus, produced with spectacle lenses, their eyes grow in axial length leading to myopia (E. L. Smith, 1998). Infant eyes (including human infants) are often hyperopic at birth. This means that the axial length of the eye is too short for the refracting power of the optics, and images are brought to focus behind the retina. The resulting defocus triggers eye growth until a better match between eye length and refracting power is established, a process termed emmetropization. Under conditions in which defocus or other types of pattern deprivation persist, the eye length continues to increase beyond a proper emmetropic value and the eye then becomes myopic. This means that the axial length is too long for the eye's refracting power.

Emmetropization is a self-regulatory process for matching eye length to refracting power that relies on some form of feedback from retinal image evaluation to scleral mechanisms for controlling eye growth. While it is reasonable to expect that perceptual signals, passing through the visual cortex, would play a role in this feedback loop, there is evidence that emmetropization can proceed even when the optic nerve is cut (Troilo, Gottlieb, & Wallman, 1987). This surprising result implies that the mechanism of emmetropization is local to the retina, and not tied directly to perceptual detection of defocus blur. It is possible that small amounts of defocus in reading, even if not perceptually noticeable, could trigger retinal mechanisms of eye growth.

Wallman and Winawer's (2004) reading hypothesis is that defocus associated with reading triggers eye growth, contributing to myopia. In addition to extended periods of defocus in central vision, due to under-accommodation during reading, these authors pointed out that peripheral vision may also be deprived of its typical

---

[15]Campbell used small (10 min-arc) black disks, viewed against a white background, to determine the amount of dioptric defocus before a subject detected stimulus blur. He studied the impact of pupil size, illumination, contrast and color on this threshold. Under optimal viewing conditions and a 3 mm pupil, the blur associated with 0.3 D blur could be detected.

range of pattern stimulation during reading. As we discussed in chapter 3, only a few letters near fixation (letters within the visual span) are resolvable during reading; most of a page of text stimulates peripheral vision with unrecognizable visual patterns. It is plausible to suggest that peripheral vision, at least that portion viewing the page of text, is in a state of form deprivation during reading. If peripheral form deprivation triggers an emmetropization response[16] (i.e., eye growth), then both defocus in central vision and form deprivation of peripheral vision during reading could contribute to myopia.

Wallman and Winawer's (2004) hypothesis may explain the puzzling phenomenon of school myopia: the increase in myopia among school children in the age range of about 8 to 12 years. Wallman and Winawer (2004) summarized three kinds of evidence pointing to a causal link between reading and an increase in myopia: (a) evidence from many countries for an association between educational attainment and prevalence of myopia; (b) a high incidence of myopia among young adults engaged in training for professions with intensive reading and other near tasks, such as law, medicine, or engineering; and (c) an increase in myopia in native American and Inuit cultures after the introduction of widespread Western-style schooling. Wallman and Winawer's (2004) reading hypothesis accounts for school myopia by proposing that the increased reading activity in the early school years confronts the visual system with text stimuli that initiate an emmetropization response, causing eyes to increase in length and to become myopic.

Sampled Displays and Low Vision

We also measured critical sample densities for several low-vision participants (R2, 1985, Figs. 5 and 6). In most cases, the critical values were equal to or less than the values for normally sighted participants at the corresponding character size. From our low-vision data, we estimated that text displays for low vision could be safely designed with as few as 13 × 13 samples per character.

Inspired by our early findings on CSD and critical window size (R1, 1985; R2, 1985), Denis Pelli led our research team in the development of a fiberscope low-vision reading magnifier (R3, 1985). This device is pictured in Figures 1, 2, and 3 of R3 (1985). The key component is the image guide, a flexible bundle of 125 × 125 plastic optical fibers, about two feet in length. The fibers in the image guide have the same spatial arrangement at the two ends so that an image formed on one end can be transmitted to the other end.[17] An objective lens, mounted in a stand at an appropriate distance from the page was attached to one end of the image guide, and an eyepiece was attached to the other end. With appropriate choice of lens powers for the object

---

[16]Smith et al. (2005) have shown that form deprivation of peripheral vision in infant monkeys can lead to an emmetropization response.

[17]The most familiar application of fiberscopes is as flexible endoscopes used in viewing internal organs through body openings. In endoscopes, one set of optical fibers, the light guide, carries light into the viewing space to provide a source of illumination. An image guide carries an image of the internal space to an eyepiece or camera.

lens and eyepiece, the fiberscope had magnifications ranging from 5X to 40X.[18] Each fiber averages the light in its small cross-section of the image, and transmits the light to its other end by means of multiple total internal reflections. The honeycomb appearance of the image (R3, Fig. 2A) shows the light samples associated with each fiber. When a fiber is broken, there is a corresponding dark spot in the image. Because the major cost of the fiberscope depended on the number of fibers in the image guide, we designed the fiberscope to just meet the minimum window size and sampling requirements for reading. An objective lens is chosen so that about four or five letters from a line of text can be seen at one time through the fiberscope. Given that the entire image is based on 125×125 fibers, this means that each letter is represented by more than 20×20 fibers (samples), enough to meet the sampling requirements for normal and low vision reading.

To read text with the fiberscope, the user moves the stand containing the objective lens along the line of text and looks into the eyepiece to see the magnified image. The eyepiece can be hand-held, mounted in spectacle frames, or mounted in a stand. Think of the fiberscope as a flexible microscope that carries a magnified image of text from the page to the reader's eye. We evaluated fiberscope reading performance on a group of 12 low-vision participants (R3, 1985), and compared their speeds to reading with a closed-circuit TV (CCTV) magnifier. All the participants could read with the fiberscope (speeds ranged from 12 to 95 wpm) with speeds that averaged 69% of their CCTV rates.

Our goals in designing the fiberscope were to produce a portable, inexpensive low-vision magnifier with very high power, meeting the sampling and window-size requirements of reading. There was a need for such a magnifier. Traditional, hand-held optical magnifiers are hard to use when the power exceeds about 6X. This is because a high-power magnifier must be held at a short working distance from the page, and the eye must be close to and carefully aligned with the magnifier's eyepiece. The result is postural discomfort, demand on manual dexterity, and difficulty in getting light onto the page. Higher powers can be achieved with CCTVs, but CCTVs are expensive and, until recently, not portable. Our fiberscope was portable (it could be folded and carried in a purse or backpack), had its own light source, and could be used with a comfortable sitting posture.

Our fiberscope magnifier was demonstrably useful for low-vision reading, but ultimately failed as a practical device for two reasons. First, it turned out to be costly to produce. A commercial company designed and built a rugged and attractive prototype,[19] but their estimated sales price was too high. Second, the fiberscope was not easy enough to use. It typically required two hands, one to move the objective over

---

[18] American manufacturers usually define the power of a magnifier as the ratio of image size through the device to image size for free viewing at a distance of 25 cm. For instance, 1.5 mm print, viewed from 25 cm, subtends 0.34°. Viewed through a 40X magnifier, the letters should subtend 13.6°. The actual magnification produced by optical magnifiers depends on usage, including the object-to magnifier distance and the eye-to-magnifier distance. For a discussion, see Bailey et al. (1994).

[19] Although the company's prototype looked nicer than our laboratory prototype, it was less functional. The objective assembly was so bulky that it could not be moved into the crevices made by the bindings of thick books, and, as a result, went out of focus for print within these crevices. This is an example of the frequent mismatch in assistive technology between the engineering solution and the user's needs.

the page and one to hold the eyepiece. Sometimes, the orientation of objective and eyepiece were misaligned, resulting in misorientation of the image. It is possible that these practical limitations could have been overcome had the potential cost for fiberscope magnifiers been low enough to anticipate a significant market.

Subsequently, Peli and Siegmund (1995) took a different approach in designing a fiber-optic low-vision stand magnifier. Their magnifier consisted of a rigid bundle of coherently arranged glass fibers, with each fiber tapered to have a smaller diameter at one end than the other. The narrower face of the tapered bundle, typically one inch in diameter, sits directly on the page. The user views the wider upper face. Magnification, determined by the extent of tapering of the fibers, is in the range of 2X to 3X. There are no lenses involved; the fiber bundle simply acts as an image conduit whose tapered arrangement produces a magnified real image on the upper surface. The number of letters visible in the field is determined by the number of letters spanning the narrower, lower face of the fiber bundle. For newsprint (character size of 1 M unit), approximately 16 letters would be imaged by a one-inch-diameter taper. If the taper has 300 fibers per inch (80 micron fibers), the sample density would be nearly 20 samples per character width, quite adequate for reading. Tests of reading performance demonstrated that normally sighted and low-vision participants could read with these magnifiers at speeds of about 60% of their maximum reading speeds (E. M. Fine et al., 1996).

There are two major differences between our fiberscope magnifier (R3, 1985) and the taper design (Peli & Siegmund, 1995): (a) We used the fiber-optic bundle as a flexible, unit magnification image conduit and relied on lenses for magnification, while the tapered bundle has "built-in" magnification without the need for lenses; and (b) our design could achieve very high magnifications (up to 40X) whereas the taper design is only practical for relatively small magnifications up to about 3X.

With the advent of personal computers in the early 1980s, the issue of image distortions from coarse sampling reemerged in the context of computer displays for low vision. Screen magnification software, such as ZoomText (AISQUARED, Manchester Center, Vermont), can magnify text by at least 16X and has facilitated computer access by many people with moderate and severe low vision. How does magnification interact with screen resolution? In one of the examples in Table 4.2 above, the x-height of 12-pt Times Roman letters on a display with 800 × 600 pixels is only 6 pixels. At a normal viewing distance of 16 inches, the letters subtend 0.36°. As we discussed above, smaller letters have lower critical sample densities, so the x-height of 6 pixels is probably not a limitation on normal reading performance. Now imagine that software magnification enlarges these characters ten times to an x-height of 3.6°. If magnification is accomplished by pixel replication,[20] the underlying sampling matrix will also be magnified so that there are still only 6 samples per x-height. To a normally

---

[20]Magnification by pixel replication involves replacing an image pixel with a group of pixels with the same gray level. For example, 10X magnification would involve replacing each pixel from a portion of the original image with 10 × 10 pixels (i.e., 100 pixels) of the same gray level in the magnified image. If the display has a total of 800 × 600 pixels, it would have room to magnify a region of only 80 × 60 pixels from the original display.

sighted reader, the resulting letters will look distorted, and diagonal strokes will have obvious jaggies. Based on our findings in R1 (1985), a sample density of 6x6 samples per letter for 3.6° letters falls below the CSD, so reading speed would be expected to slow down.

The practical question, however, is whether the jaggies adversely affect low-vision reading speed. Bailey, Boyd, Boyd, and Clark (1987) timed participants as they searched for a target letter in an array of other letters. The letters were rendered in the Times Roman font with either 12 or 24 pixels per letter height. When the letters were magnified to twice the acuity limit, the participants performed the search task faster with the higher-resolution letters. Although potentially relevant to text reading, we cannot be sure that the sampling limitations on visual search are the same as for reading.

Li, Nugent, and Peli (2000) studied the effect of jaggies on letter recognition in normal peripheral vision. They were interested in this issue because of its relevance to low-vision reading with central-field loss. Rather than manipulating sampling density, they compared performance for two types of rendering: binary letters, composed of black and white samples, on a 16×17 matrix, and the same letters with grayscale smoothing ("antialiasing"). Ten letters were used from the Helvetica font, with angular size of 0.6°. Although the underlying resolution of the two types of letters was the same, the jaggies were much more salient in the binary letters. The participant's task was to recognize a pair of letters flashed briefly, one at fixation and one in peripheral vision (from 2.5° to 12.5° eccentricity.) As expected, letter-recognition performance declined with increasing eccentricity, but there was no difference between the binary and grayscale letters. This study did not use text reading, but even if it had, we would predict no difference in the two conditions from the results of R1 (1985). This is because the effective sample density in the Li et al. study is roughly 16×17, which our data indicate is above the CSD for 0.6° characters. For this reason, we would not expect distortions due to coarse sampling to affect reading speed, and we would not expect any additional benefit from grayscale smoothing.

These studies do not entirely resolve whether jaggies are an important consideration for computer displays for low vision. Nevertheless, prudent design of screen magnification software for low vision would include some form of antialiasing in addition to magnification by pixel replication.

An interesting application of display sampling concerns the design of retinal or cortical stimulating arrays for prosthetic vision devices. This topic is discussed in Box 4.2.

Box 4.2
Pixelized Vision as a Prosthetic for Blind People

Most forms of vision loss are attributable to disease of the retina or optics. What about bypassing the retinal and preretinal stages and directly stimulating the visual cortex? Brindley and Lewin (1968) tested this idea on a 52-year-old nurse who was almost totally blind in both eyes from glaucoma. They implanted a rectangular lattice of 80 electrodes intracranially to contact the surface of her visual cortex (occipital pole of the right hemisphere). These electrodes were connected by cable to a set of 80 radio receivers, encapsulated in rubber, beneath the paracranium and secured by screws to the skull. Radio signals could be sent selectively to one of the receivers which in turn initiated an electrical stimulus directly to the visual cortex. The participant perceived points or clouds of light (phosphenes) in her left visual field. Simultaneous stimulation with several electrodes yielded percepts of simple pattern configurations. Brindley and Lewin concluded that direct cortical stimulation could be used as a visual prosthesis for blind people. They estimated that between 50 and 600 properly arranged stimulators would be all that is necessary for functionally useful reading.

In subsequent years, several research groups have pursued the quest of developing a visual prosthesis, based on direct cortical stimulation. There have been many technical impediments—the general problem of biocompatibility of neural implants, fabrication of stimulating electrode arrays, lack of suitable animal models and psychophysical evaluation methods, and the invasive nature of brain surgery for inserting an implant. The recent development of safe and effective microelectrode arrays for neural prosthetics is a promising step (cf. Normann, Maynard, Rousche, & Warren, 1999). A complicating factor is the non-conformal map of the visual field onto the visual cortex, creating the need for an inverse mapping algorithm that connects the location of each electrode to that of a phosphene in visual space. There is the additional important question whether the visual cortex of people with long-term blindness remains capable of processing visual inputs in the normal way. For a review of this issue, see Merabet, Rizzo, Amedi, Somers, and Pascual-Leone (2005).

In recent years, the idea of basing a visual prosthesis on direct stimulation of visual neurons has blossomed in the form of retinal prosthetics rather than cortical prosthetics. Interest in this idea has been motivated by the success of cochlear implants, which have restored functionally useful hearing, including speech perception, to many people with severe hearing loss.

Cochlear implants bypass the normal process of mechanical activation of the ear's receptors and directly stimulate the spiral ganglion cells, whose axons form the auditory nerve.

There are two types of retinal implants—subretinal and epiretinal. The subretinal implants are placed between the retinal pigment epithelium and the inner nuclear layer of the retina, in the location of the degenerated photoreceptors. The implant contains light-sensitive microphotodiodes equipped with tiny electrodes. When sufficient light is captured, the diodes generate currents that stimulate retinal neurons. We can think of the diodes as artificial photoreceptors, replacing the damaged biological photoreceptors. Epi-retinal implants are placed on the ganglion-cell side of the retina. The implant consists of a micro-electrode array for stimulating bipolar or ganglion cell bodies. The microelectrode array receives image data from a camera, perhaps mounted on glasses, and converts the image to a pattern of electrical stimulation. Both approaches rely on functioning retinal ganglion cells and optic nerve. They are both intended to help people with diseases primarily affecting the photoreceptor layer of the retina, such as retinitis pigmentosa and age-related macular degeneration. For discussion of the pros and cons of the two designs, see Zrenner (2002).

There is an important difference between the two designs that might have an impact on reading. In the epi-retinal design, image capture is accomplished by a head-borne camera. Head movements would be necessary to look around, including scanning of a line of text, while eye movements have no effect, since the electrodes are stabilized on the retina. In the subretinal design, image capture is accomplished by the light-sensitive diodes on the implant attached to the retina, so eye movements, rather than head movements, could be used in scanning a line of text.

Both types of retinal implants, and also the cortical implants, deliver coarsely sampled input to the visual system through stimulation of neurons at a few spatially separated sites. Studies with patients using epiretinal stimulation have employed electrode separations equivalent to roughly 1.5° to 2.7°, and have determined that the resulting phosphenes are perceptually discriminable (Humayun & de Juan, 1998; Humayun et al., 2003).[i] Although the details of the percepts associated with individual electrode characteristics (placement, current level and frequency) may vary quite widely, we can imagine the overall perceptual effect to consist of a somewhat irregular array of bright spots in the visual field, a perceptual experience distinctly different from normal vision.

This type of visual image is sometimes termed "pixelized vision," referring to the discrete "picture elements" associated with the stimulating electrodes.[ii] The portion of the visual field covered by the pixelized image will depend upon the size of the stimulator array and the corresponding area of stimulated retina (or cortex). Figure. 4.6 is a simulation providing some intuition for the information available from a static view of text with a 25 × 25 retinal prosthesis.

A critical issue for visual prosthetics is the number of stimulating electrodes necessary for useful vision. While any vision, no matter how coarse, is likely to be of functional benefit, a demonstration of function-

---

[i] In the Humayun et al. (2003) study, discussed in a later paragraph of Box 4.2, the center-to-center separation of adjacent electrodes in a 4 × 4 array was 720 μm. Since one degree of visual field is imaged on about 270 μm on the retina, the electrode separation corresponds to about 2.7° in the visual field.

[ii] Perception from visual prosthetics has also been termed "artificial vision." This term may be a misnomer. Although the source of the stimulus is unusual and perhaps artificial in the case of implants, the resulting visual percepts depend on the biology of the visual pathway. It may be preferable to reserve "artificial vision" for computer-based vision.

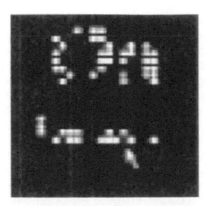

Figure 4.6   Simulation of text as seen with a retinal prosthesis. This figure illustrates the information that might be available in a static text view with a 25
× 25 retinal prosthesis. Portions of two lines are visible, including a part of
the word "one" opening the sentence "One time a blue bird built a nest ... ."
The display contains some missing samples, simulating the likely occurrence of dropouts in a real prosthetic display (figure courtesy of Gislin
Dagnelie).

ally useful print reading with a visual prosthetic would be a cause for cel-
ebration. Brindley and Lewin (1968) believed that between 50 and 600
stimulators would be necessary for fluent reading. Our study of sampling in R1, 1985) indicated that regular arrays of between 4 × 4 (16)
samples and 20 × 20 (400) samples per letter are required for rapid
reading, depending on angular character size. It is unclear if the character-size dependence of critical sample density applies to pixelized vision. If it does, and if the samples have a separation of roughly 1.5°, our
data indicate that 8 × 8 (64) samples per character would be adequate.
In Section 4.4 below, we will see that about five characters must be visible at a time for rapid reading. From these figures, we estimate that approximately 320 samples (5 characters, each represented by roughly 64
samples) would be minimally necessary for fluent reading.[iii] These
rough calculations imply modest sampling requirements for reading
and make it plausible to hope that a retinal prosthesis, based on a few
hundred stimulators, could provide useful reading vision. Of course,
only empirical research on patients can tell us how effective the implants
will be for reading.

Simulation studies by Sommerhalder et al. (2003; Sommerhalder,
Rappaz, & de Haller, 2004) help bridge the gap between our original
sampling study and the likely outcome with retinal prosthetics. Normally
sighted participants viewed pixelized images (similar to the sampled

---

[iii]But findings to be discussed in section 4.4 below indicate that additional pixels would
likely be necessary to support page navigation in realistic reading situations.

stimuli studied in R1) on a computer screen. Individual four-letter words were presented for recognition at retinal eccentricities from 0° (fovea) to 20° in the lower visual field. The number of samples in the pixelized images varied from 83 (approximately 15.4 × 5.4) to 875 (approximately 50 × 17.5). The images were scaled in size in peripheral vision. For instance, at 15° eccentricity, the images subtended 10° horizontally by 3.5° vertically (roughly comparable to the region that would be stimulated by a 3 mm × 1 mm retinal implant). An eye tracker was used to monitor eye movements and stabilize images on the retina, simulating the fixed location in the visual field of input from a retinal implant.

In foveal vision, the participants achieved high accuracy in word recognition for images with 286 pixels (28.6 × 10), and performance deteriorated quickly for coarser images. Although overall accuracy was poorer in peripheral vision, there was the same dependence on the number of samples, with performance dropping quickly below 300 pixels. There is good congruence between Sommerhalder et al.'s (2003) finding of a 300-pixel requirement and the above estimate of a 320-sample requirement from our study (R1, 1985). Sommerhalder et al.'s (2003) study extends the findings to more realistic simulations of prosthetic vision by testing peripheral vision with image stabilization.

In a second simulation study, Sommerhalder et al. (2004) extended their methods to measure reading performance for continuous text. Normally sighted participants read text through a 10° × 7° viewing window, stabilized on the retina at 15° in the lower visual field. The window contained a pixelized image with 600 pixels, roughly six letters wide and two lines high. The participants could move through the text with eye movements, simulating a sub-retinal prosthetic. The participants received extensive training in the reading task, about 1 hr a day for 2 months. Initially, reading performance was very poor for two of the three participants. After training, all three participants could read text accurately at speeds between 14 and 28 words per minute. An important part of the training involved suppression of reflexive saccades that attempted to bring the stabilized peripheral image to the fovea.

The results of these simulation studies are important in showing that reading with a retinal prosthetic should be possible if there are at least 300 to 600 stimulators. The results also alert us that learning to read with a retinal prosthetic will require extensive training and may not come close to restoring normal reading speed.

Of course, retinal prosthetics could be useful for many visual functions besides print reading. Studies of coarse pixelized vision, simulating retinal implants, have evaluated face recognition (Thompson, Barnett, Humayun, & Dagnelie, 2003) and simple object-recognition tasks and manual tasks, such as candy pouring (Hayes et al., 2003).

But what about patients with actual retinal implants? What can they do? We do not yet know. In the first detailed study, Humayun et al. (2003)

reported on a series of tests of one participant with an epi-retinal implant. A 4 × 4 platinum electrode array was successfully implanted in the temporal retina of the patient's right eye. The patient, a 74-year-old man had been blind for many years from retinitis pigmentosa. Stimulation by each of the 16 electrodes produced percepts of localized spots of light. For the most part, the relative positions of the phosphenes matched those of the stimulating electrodes. When the array was connected to a camera, the patient was able to localized sources of light in the room, detect motion, and locate a large, dark object in the field. But this pixelized vision was too coarse to attempt reading.

Future research will reveal the practical limits on the number of stimulators in retinal implants, and hence the limits on pixelized vision. It remains an open question how much visual function these displays can support, and whether print reading will be possible.

4.4 THE WINDOW-SIZE EFFECT AND PAGE NAVIGATION

In the design and prescription of low-vision magnifiers, there is often a trade-off between the angular size of characters (magnification) and field size. See Figure 4.7. Consider, for example, a display where the field size is limited to 15°. If the text letters average 1° in width, there will be approximately 15 characters (including spaces) visible at a time from a line of text. But if the characters are magnified to 5°, only about three characters from a line of text will be visible at a time. We refer to the number of characters visible on a line of text in a magnifier's field of view as the *window size*.

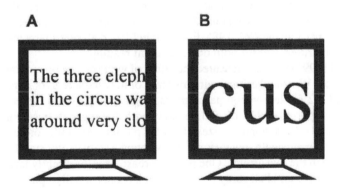

Figure 4.7 Window size refers to the number of characters from a line of text simultaneously visible in the field of a display. The two panels show displays of the same physical size. In Panel A, the window size is 15 characters. In Panel B, the characters are five times larger and the corresponding window size is three characters. If the figure is viewed from a distance equal to four times the display width, the angular display width will be about 15°.

Assuming that the whole line of text does not fit within the field of view, the reader will have to move the magnifier along the line of text or, alternatively, move the text through the field of view of a stationary magnifier. We refer to the process of moving a magnifier through text, including retrace from the end of one line to the beginning of the next, as *page navigation*. As the window size decreases (fewer letters visible at one time), there is a greater need for page navigation.

In the real world of low-vision reading, it is evident that magnification, window size, and page navigation all interact. In our laboratory research, we have tried to disentangle these factors. In section 3.2, we considered the effect of character size (magnification) on reading speed in both normal and low vision. We now turn to window size and page navigation.

How does window size, the number of characters simultaneously visible in the field, affect reading speed? In R1 (1985), we measured reading speed as a function of window size for normally sighted participants, and in R2 (1985) for low-vision participants. We used the drifting-text method (see section 2.1 in which text scrolled through the window automatically at a specified rate. This method eliminates the need for the reader to control page navigation. We produced windows with different sizes by occluding portions of the display screen with black paper. The window sizes varied from 0.25 characters (just a sliver of a character could be seen through the window at one time) to 20 characters.

The simple pattern of results is shown in Figure 4.8 (reprinted from R1, 1985, Fig. 8). Reading speed increased as window size increased up to four characters, and then quickly leveled out. For window sizes below the critical value, reading speed showed a square-root dependence on window size. For instance, for a four-fold increase in window size from one to four characters, reading speed increased by about a factor of two. Slow reading was usually possible even for window sizes less than one character. This general pattern of results held for participants with normal vision (R1, 1985) and for a heterogeneous group of low-vision participants (R2, 1985, Fig. 7). It held for different character sizes, both contrast polarities, and for both oral and silent reading.

In R14 (1996) we also measured the effect of window size on reading speed for drifting text. We found critical window sizes of 4.7 characters for normal participants and 5.2 for low-vision participants, slightly larger than the value of four characters reported in R1. In a similar study with drifting text, E. M. Fine and Peli (1996) found critical window sizes of 5.4 for participants with normal vision and 6.3 for participants with low vision.[21] They found a statistically significant difference between the two groups. One possible reason for the slightly larger window sizes in the Fine and Peli study is their use of a proportionally-spaced font, rather than the fixed-width font used in our studies. In the following discussion, we take

---

[21]Fine and Peli (1996) measured critical window sizes for sentence reading and also for reading of random strings of words. The critical window sizes cited here refer to their results for sentence reading. The values of critical window size from R14 (1996) are for a criterion of 85% of maximum reading speed, while the figures from Fine and Peli (1996) are for an 80% criterion. Criterion effects are discussed further in the text.

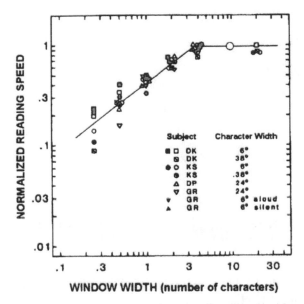

Figure 4.8 Effects of window width on reading rate. Normalized reading rate is plotted as a function of window width (measured in number of character spaces) for several experiments. The large symbol at a window width of 10 characters indicates that normalization was computed with respect to the reading rate for this condition. The data have been fit with lines having slopes of 1/2 and zero in these log-log coordinates. Data are shown for four observers, and for white-on-black text (open symbols) and black-on-white text (solid symbols). The solid triangles compare silent and oral reading rates. (From R1, 1985, Fig. 8.)

the critical window size for drifting text to be about five characters, recognizing that empirical estimates vary slightly on both sides of this value.

The critical window size of about five characters implies that a rather small field of view can support fast reading. Subsequent research in other laboratories, using different methods, yielded discrepant findings implying the need for larger windows. Whittaker and Lovie-Kitchin (1993) made an observation that suggested a resolution to the discrepancies; the critical window size seemed to be smaller in experiments in which participants read text that automatically drifted across a TV screen than in experiments in which participants manually move the text through the field of view of a CCTV magnifier.

A CCTV magnifier (Fig. 4.9) consists of a display monitor and a video camera equipped with a zoom lens. The camera and lens are mounted above a movable platform. Printed material is placed on the platform. The user views the magnified image on the monitor screen, and navigates through the text by manually moving the platform. It is quite feasible to achieve 40 times magnification with a CCTV (i.e., the ratio of x-height on the monitor to x-height on the printed page, both mea-

Figure 4.9   Closed-circuit TV (CCTV) magnifier.

sured in cm, is 40:1.). For a discussion of all aspects of CCTV magnifiers, see Lund and Watson (1997).

Page navigation with a CCTV monitor involves learning an unfamiliar visual-motor skill. To read a line of text, the user moves the platform from right to left through the camera's field of view. At the end of the line, the user retraces to the beginning of the new line by moving the platform from left to right.[22] It is plausible that the combined visual and motor demands of this form of page navigation might benefit from a grater field of view than reading in the absence of these navigational demands.

In R14 (1996), we considered the interacting effects of page navigation and window size. We compared reading speeds for two conditions of page navigation—automatically drifting text (no manual page navigation required), and manual navigation of a CCTV platform—using the same video display. We replicated the finding from our earlier study of a small critical window size for drifting text of about five characters. But in confirmation of the observation by Whittaker and Lovie-Kitchin (1993), we also found that the critical window size

---

[22]These manual movements for reading with a CCTV magnifier reverse the directionality of eye movements in reading, or manual movements in reading with a hand-held magnifier. The reversal of direction occurs because the text is moved through the field of a stationary camera rather than moving the camera along the line of text. The unfamiliar pattern of motor movements in reading with a CCTV requires a process of motor learning by a new user. A different pattern of motor movements is required for reading magnified text on a computer screen. With screen magnification software, such as ZoomText, the user typically moves the magnified window along a line of text by moving the mouse. Standard left-to-right reading of the line is associated with left-to-right mouse movements, simulating the movement of a hand-held magnifier along a line of text, and opposite to the pattern of manual movements with a CCTV. People who use both a CCTV magnifier for hardcopy text and a software magnifier for computer text must learn to use opposite patterns of manual movements for page navigation in these two cases.

was larger when the participant was responsible for manual movements of the CCTV platform—the critical window size increased to 8 characters for low-vision participants and to 12 characters for normally sighted participants.

We identified a second factor contributing to the discrepancies in the literature on estimates of the critical window size. For drifting text (no manual page navigation), the curve of reading speed versus window size shows a fairly sharp transition near 5 characters; the curve is nearly flat for larger characters. For manual page navigation, there is a softer transition of reading speed from a strong dependence to a weak dependence as window size increases. This being the case, a criterion reading speed needs to be picked for defining the critical window size. In R14 (1996), we showed that the selection of different criteria (e.g., window sizes yielding 85% of maximum reading speed, 65% of maximum reading speed, ...) had substantial effects on the numerical estimate of critical window size.

Taking these two factors into account—automatic text presentation versus manual page navigation, and the nature of the performance criterion for defining critical window size—we were able to reconcile the discrepant findings on window-size effects. In R14 (1996, Fig. 9), we demonstrated the close agreement of speed versus window-size curves for three studies all requiring manual page navigation (R14, 1996; Lovie-Kitchin & Woo, 1988; Lowe & Drasdo, 1990).

Most of the research on window-size effects has been done with video magnifiers, but the results may generalize to optical magnifiers. A. R. Bowers (2000) showed that page navigation with hand-held and spectacle-mounted optical magnifiers yields the same reading speeds for window sizes from 7 to 29 characters, implying a window-size requirement of no more than about seven characters. E. M. Fine et al. (1996) showed that reading speeds measured with a stand magnifier (the fiber-optic taper described in section 4.3 above) increased rapidly for window sizes from one to five characters, and then much more slowly for window sizes up to 13 characters.

Our current understanding is that for drifting text, not requiring any manual page navigation by the participant, the critical window size for reading is about five characters. But when the participant is obliged to move the magnifier over the text or scan the text through the magnifier's field of view, the requirements increase to 8 or more characters (for 85% of maximum speed. For CCTV, reading speed continues to show a slow rise as window size grows up to about 20 characters.

Relation to Visual Span

In this subsection, we discuss the relationship between window size and visual span, and consider two models for explaining the critical window size of five characters.

In section 3.7, we reviewed evidence for the hypothesis that the visual span is a sensory bottleneck on reading. The visual span refers to the limited number of letters that can be recognized with high accuracy during each fixation in reading. We represent the visual span empirically by visual-span profiles, plots of Letter-recognition accuracy versus letter position left or right of fixation. For an example, see the lower panel of Figure 3.8. For normally sighted readers, the visual span is approximately 10 characters in size, that is, letter recognition accuracy is moderately high for four or five characters right and left of fixation, with substantially decreasing accuracy further out from fixation.

What is the relationship between window-size measurements and visual-span measurements? The major difference is that the window-size effect refers to the size of the stimulus display, while the visual span refers to letter-recognition capacities of the reader's visual field. The distinction is analogous to the difference between the field size of a magnifier and a patient's visual field measured with a perimeter. In the following paragraphs, we outline a way of linking the visual span to the window-size effect in reading drifting text. We propose a way of reconciling the critical window size of five characters (in the absence of page navigation) and the normal visual span of 10 characters.

Figure 4.10 is a diagram showing how the normal visual span relates to drifting text. Suppose a line of text emerges at the right margin of the screen and scrolls at constant speed from right to left. Initially, the eyes fixate on the leading letter and track it across the screen with smooth-pursuit eye movements, while subsequent letters continue to emerge at the edge of the screen. Eventually, the reader makes a saccade back to the right edge of the screen to fixate on a new letter, and the process

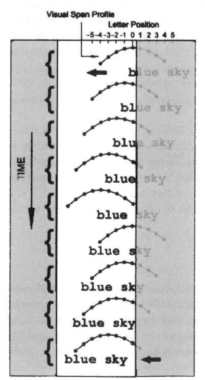

repeats. This general pattern of eye movements has been observed in the reading of drifting text (R1, 1985; Buettner et al., 1985). How far do the eyes track the fixated letters? How many letters emerge onto the screen before the eyes saccade back to make a new fixation? Once the window size exceeds this saccade size, it is unlikely that further increases in window size will affect reading speed.

Figure 4.10   Diagram of visual span and drifting text. As the words "blue sky" drift into view at the right margin of the screen, the eyes track the leading letter "b" across the display so that the visual span remains centered on b. Eventually, the eyes saccade back to the right margin of the display and the peak of the visual span becomes aligned with the "s" of sky. Shaded regions indicate information off screen and not seen by the reader.

We can translate the drifting-text process into the visual-span framework as illustrated in Figure 4.10. The fixated letter lies at the center of the fovea in "Slot 0" of the visual span. As the eyes track this letter across the screen, a new letter scrolls onto the screen and occupies Slot 1. Then a third letter appears on the screen and occupies Slot 2 of the visual span, and so forth. Smooth pursuit of the scrolling text results in sequential stimulation of letter slots in the right half visual span, starting with Slot 0, then Slot 1, then Slot 2.

How long does the participant fixate the leading letter? In other words, how many letters drift onto the screen during the fixation before the reader saccades back to fixate a new letter at the right margin? Two opposing factors may govern the decision about when to make a saccade. First, as the reader tracks the leading letter further to the left, newly emerging letters at the right margin of the screen occupy visual-span slots further out on the right side of the visual-span profile. These slots provide less and less reliable information for letter recognition. There will be pressure to make a saccade as the newly appearing letters become less and less informative. Second, there is a time requirement for programming and executing a saccade. If saccades were instantaneous and required no effort, the reader would probably make a saccade to fixate each new letter as it appeared at the right margin of the screen, bringing each new letter into Slot 0 of the visual span. (Typically, highest letter recognition accuracy occurs in Slot 0.) If this were the case, the critical window size for reading drifting text would be only about one letter. But because there is a time requirement (latency) for making saccades, smooth pursuit is prolonged, even though the newly appearing letters are recognized with reduced accuracy further to the right on the visual-span profile.

The question is how the visual system balances these two constraints—the need to prolong smooth pursuit while preparing for the next saccade, and the decreasing informativeness of letters further out on the visual-span profile. Suppose there is a minimum of 200 msec between saccades in the drifting-text paradigm.[23] The drift rate will determine how many letters appear on the screen between saccades, and the corresponding number of slots in the right visual span used for letter recognition. For instance, if the scroll rate is 250 wpm, five letters will drift into view in 200 msec, and if the scroll rate is 500 wpm, 10 letters will appear[24] in 200 msec. Because the first five letters will fall into Slots 0, 1, 2, 3, and 4 of the visual span, for which accuracy is fairly high, we would expect reading performance to be good at a scroll rate of 250 wpm. But at the higher scroll rate of 500 wpm, letters 6 to 10 on each fixation would be processed by low-accuracy slots on the tail of the visual span profile, and we would expect reading errors.

---

[23]This value is close to the minimum latency for saccades in reading, estimated to be 150 to 175 ms (Rayner, 1998).

[24]Adopting the arbitrary estimate that each word of text and the following space occupy an average of six characters (Carver, 1990), 250/minute is equivalent to 1500 characters/minute, also equivalent to 25 characters/sec. At a drift rate of 25 characters/sec, one character appears each 40 ms, and five characters appear in 200 ms.

In R1 (1985), we observed that normally sighted participants could read drifting text accurately up to a maximum scrolling rate of about 250 wpm. For higher scrolling rates, participants made a substantial number of errors. We have also determined that the critical window size is close to five characters. The interpretation given here is that the joint constraints of saccade latency and decreasing letter recognition on the visual-span profile account for the maximum reading rate and for the critical window size. Put in simplest terms, the critical window size of about five characters is mainly a consequence of the use of only about five high-accuracy slots in the visual span (the central slot and 4 to the right) for reading drifting text.

The foregoing account ties the critical window size to the half-width of the visual span. Accordingly, we would predict that the critical window size should covary with the size of the visual span. One way of testing this prediction would be to measure the critical window size for normally sighted participants under conditions in which the visual span is known to shrink, such as near the contrast threshold or near the acuity limit. The most straightforward prediction is that the critical window size should be smaller when the visual span is smaller.

The above explanation for the window-size effect faces a major unresolved challenge. If it is true, as argued in chapter 3, that low-vision participants have smaller visual spans than normal participants, we would expect that their critical window sizes would also be smaller. In R2 (1985), we measured the effect of window size on reading speed for low-vision participants. Although the low-vision reading speeds were slower, we did not find that the critical-window sizes were systematically smaller. This result could indicate a problem with the proposed tie between the visual span and the critical window size, or it might indicate that some other factors influence the critical window size in low vision.

In R1 (1985), we did not offer an explanation for the critical window-size of four characters, but we did note that this value was very close to the mean word length of 4.1 characters in our text samples. A variant of the model just proposed could link window size to word length rather than to visual span. As above, suppose the reader fixates the leading letter of a word as it scrolls into view at the right margin. The reader tracks the letter's movement across the screen while the remaining letters of the word scroll into view. Smooth tracking continues until a space is detected, indicating the end of a word. Detection of the space triggers a saccade back to the right edge of the screen where the eyes pick up the leading letter of the next word. If this is the mechanism controlling smooth pursuit and saccades in drifting text, reading performance should be good as long as word length does not exceed window size. If the word is too long (e.g., an 8-character word drifting through a 4-character window), an adjustment would be necessary; either the reader would need to make an intraword saccade, or infer the word from the leading four letters. Presumably, these adjustments could result in performance errors. Further assumptions would be necessary to make a quantitative estimate of the effect of window size on reading speed for this model, but there is a clear prediction. The critical window size should depend on the frequency distribution of word length in text. In the simplest case, we would expect the critical window size to be correlated with mean word length.

In this subsection, we have considered two explanations for the critical window size of about five characters for reading drifting text. Consistent with eye tracking data, both explanations assume that the eyes fixate a character in text as its scrolls across the screen, and periodically saccade back to the right margin to fixate a newly appearing character. The critical window size is determined by the distance over which the eyes track the scrolling text before a saccade occurs. The models differ in the factors triggering the saccade—either the decline in letter-recognition accuracy at increasing distance from the fixated letter, or the detection of a word boundary.

4.5 NAVIGATING THROUGH TEXT

Readers process prodigious amounts of text in a lifetime. Pelli et al. (in press) estimated that people who read for one hour per day for 50 years recognize about 1.4 billion letters. Research over the past 125 years, reviewed in Box 3.1, has shown that the visual system processes text in small chunks of only a few letters at a time. Of fundamental importance to reading is the selection of a mechanism for moving efficiently through text from chunk to chunk. We refer to this process generically as page navigation.

Eye Movements in Reading

Everyday reading capitalizes on the speed and accuracy of saccadic eye movements for page navigation. There is a vast literature on eye movements in reading, dealing with the oculomotor processes per se, and the implications for related cognitive processes. Reading eye movements have been described in detail in many sources reviewed by Rayner (1998), Rayner and Pollatsek (1989), and Just and Carpenter (1987). For a general treatment of eye movements, see Carpenter (1988).

Reading is characterized by saccades, averaging 8 characters in length, separated by fixations, averaging 225 msec in duration (Rayner, 1998). Both the length of saccades and the duration of fixations are under the online active control of the reader. Is this ongoing cycle of saccades and fixations necessary for rapid reading and good comprehension, or can reading adapt effectively to other types of page navigation?

Eye Movements Are Not Necessary. Although eye movements are by far the most common vehicle for page navigation, they are not necessary for fast and effective reading. Gilchrist, Brown, and Findlay (1997) described a woman with ophthalmoplegia who was unable to make eye movements. Despite this disability, her reading was functionally normal; she relied on a series of saccades separated by fixations. But unlike normal reading, she made head saccades (rapid head movements) rather than eye saccades. The head saccades averaged six or seven characters in length, separated by 200 msec, and her average reading speed was about 250 wpm.

Saccades Are Not Necessary.   Saccades *per se*, whether based on eye movements or head movements, are not necessary for reading. RSVP is a method for presenting text which minimizes or eliminates the use of eye movements (see section 2.1). It is possible to read rapidly and with good comprehension with RSVP. Juola, Ward, and McNamara (1982) tested participants for comprehension on paragraphs of text, presented on a computer display in either a standard page layout or in RSVP. They varied the RSVP exposure rate so that the overall reading time bracketed the page reading time. There were no important differences in comprehension between the two display formats, even when the overall RSVP exposure time was half that for page reading (that is, RSVP reading speed was twice that for page reading).

Active Control Of Fixation Times And Saccade Lengths Are Not Necessary. RSVP equalizes the exposure time per word, eliminating the opportunity for readers to fixate longer on some words than others. Nevertheless, as just described, RSVP supports rapid reading, and comprehension remains comparable to conventional page reading.

Bouma and de Voogd (1974) asked whether the reader's active control of saccade length is critical for rapid reading. They studied oral and silent reading speed in a procedure in which lines of text advanced past the point of fixation in steps of a set length ("line-step" method). The length of these steps was a parameter, and simulated the effect of saccades of a given length. Bouma and de Voogd's method also treated the pause time between steps as a parameter, simulating fixation times in reading. With suitable choices of pause times and step lengths, their participants were able to achieve oral reading rates between 250 and 310 wpm with the line-step presentation, and substantially faster speeds for silent reading. Bouma and de Voogd concluded that rapid reading is possible with steady fixation, and externally controlled pause times and step lengths. Taking these results together with those from RSVP, we can conclude that active control of fixation times and saccade lengths is not necessary for rapid reading.

A variant of Bouma and de Voogd's (1974) line-step method, termed "leading" is sometimes used in small electronic displays. Instead of long steps (roughly equivalent in size to saccades) and pauses roughly the length of fixations, the "leading" method uses short steps, typically one character, with very brief pauses between steps. This version of the line-step method approximates the drifting-text method, replacing smooth motion with a rapid series of one-character (or several character) steps. In contrast to smoothly drifting text, Leading is jerky in appearance, and has been reported to yield poor reading performance (Juola, Tiritoglu, & Pleunis, 1995). We return to the reason for this poor performance in the subsection on Small Displays below.

Stable Fixation Is Not Necessary.   In the drifting-text method, used in our early psychophysical studies (R1, 1985) and discussed in some detail in sections 2.1 and 4.4, the eyes do not use stationary fixation at all. Instead, the reader uses smooth-pursuit eye movements to track a target letter as it drifts from right to left across the display screen. After tracking the letter through several character spaces, the eyes make a saccade to the right and begin to track another letter as it drifts from right to left. Periods of smooth pursuit from right to left are separated by saccades from left to right. The eyes are continually in motion during this type of page navigation, with no stationary fixations at all. Nevertheless, reading speeds from 250 to 350 wpm can be achieved (R1, 1985; R10, 1991), with good comprehension (R7, 1987). Successful reading of drifting text reveals that stationary fixations are not necessary for rapid reading. It should be noted, however, that the effect of smooth pursuit of drifting text is to stabilize the images of letters on the retina, that is, the eye movements compensate for and largely eliminate the retinal-image motion that would otherwise be produced by the stimulus drift. Although stable eye fixation may not be necessary for rapid reading, it is possible that stable retinal images of text (at least for brief periods) may be necessary for effective reading.

The major conclusion from the foregoing paragraphs is that human vision can flexibly and effectively adapt to a wide range of page-navigation methods. Eye movements are not necessary, nor saccades, nor even fixations. This conclusion does not diminish the importance of oculomotor factors in ordinary reading. Just as walking is the most common and widespread form of personal navigation, people also rely on a variety of alternative methods for ambulatory mobility including running, ice skating, roller blading, and the use of wheelchairs, and bicycles. The existence of these alternatives does not negate the importance of walking, but implies dissociation between the functional goal of mobility and the means for getting there. Similarly, eye movements are certainly important in reading, but other methods can be enlisted in navigating through text.

It is also useful to realize that alternative forms of page navigation can be mastered quite quickly and do not require much practice. Rapid, functionally useful reading can be achieved with minimal training using RSVP, drifting text or even Bouma and de Voogd's (1974) line-step procedure. Despite the lifetime of experience with eye-movement mediated reading, people with normal vision adapt easily to alternative forms of page navigation.

Designers of unusual display formats for normal vision, requiring one of the alternative forms of page navigation, and designers of low-vision magnifiers can take heart in the knowledge that good reading performance is possible with alternative methods for navigating through text.

Small Displays

The proliferation of electronic text displays on appliances such as microwave ovens and CD players, and portable devices, such as cell phones and notepad computers, has motivated commercial interest in small text displays and the methods for navigating through them. How small can such text displays be?

Our discussion of the window size effect in section 4.4 implies that a display size of only five characters is sufficient, provided that the text automatically drifts across the display.

Juola et al. (1995) tested participants with the line-step approximation to drifting text (termed the "leading" method). They used an eight-character display and one, two or three-character steps. For example, for the sentence "Mary had a little lamb," and a step size of one character, three successive presentations would show:

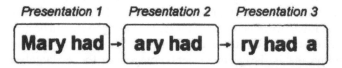

The line-step text was read poorly (only 42% of words were read correctly at presentation speeds of 171 and 260 wpm). Why is smoothly drifting text a good stimulus for reading but line-step presentation with one-character steps is poor? Suppose the line-step method is used to present text at 250 wpm with one-character steps. For simplicity, assume that 250 words and their trailing spaces occupy 1,500 characters (equivalent to assuming that the average word and its trailing space occupy six characters), then there will be 1,500 steps in 60 sec, equal to 25 steps per second, or one step in 40 msec. The individual steps are easily discriminable from continuous motion, giving the right-to-left motion of the text through the display a jerky appearance.[25] As discussed earlier in this section, drifting text is read with periods of smooth-pursuit eye movements separated by saccades. It is likely that the jerkiness of the line-step presentation is ineffective in entraining smooth-pursuit tracking. When the line-step method uses long steps (roughly equivalent to saccades) and long pauses (roughly equivalent to normal fixations), the reader need not track the right-to-left movement of the text, and can read effectively by maintaining stable fixation at a given point on the display (Bouma & de Voogd, 1974).

RSVP has also been considered for small-display applications, because only one word need be displayed at a time. Mean word length in English is only 4.7 characters, and 90.14% of words in text are eight characters or less.[26] An electronic

---

[25]In several papers in our series, we created drifting text by panning stimulus images across a pixel display. For details, see R5 (1987). Because the raster display refreshed at a 60 Hz rate, the text advanced from right to left across the screen in discrete steps, similar to the line-step presentation method. For the example of a presentation of 250 words/minute, equivalent to 25 characters per sec, the step size would be less than half a character and the pause length between steps approximately 16 ms. Text presented in this way appears to drift smoothly and is effective in entraining smooth pursuit eye movements. Presumably, the high temporal frequency (60 Hz) of the discrete steps exceeds the spatiotemporal resolution of vision.

[26]These figures were computed from the frequency distribution of word lengths in the British national Corpus (Kilgarriff, 1997).

display, capable of showing eight characters simultaneously could present most English words without difficulty. Longer words or names could be split across multiple frames using a hyphen. For instance an eight-character RSVP display could present "deoxyribonucleic acid" in four RSVP frames as follows:

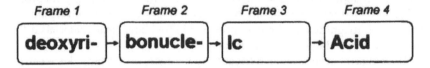

Juola et al. (1995) showed that participants could read with high accuracy (96% of words correct at a presentation rate of 260 wpm) on an eight-character RSVP display of this type.

Although people can achieve high reading speeds with RSVP, they tend not to like it. There are two reasons: First, it is a forced-march method, seeming to take control away from the reader. Second, it discards all page layout information. Efforts have been made to reduce these complaints by allowing the reader adjustable pacing, by introducing pauses at the end of sentences or other punctuation points, and by including scroll bars or other iconic indicators of one's current location in the text (cf. Castelhano & Muter, 2001; Rahman & Muter, 1999). These improvements increase user satisfaction, but are not sufficient to overcome their overall dislike of RSVP reading.

While the minimal display size for reading continuous text, based on visual constraints alone, is indeed small—one word or less—people evidently prefer larger displays for manipulating text and for utilizing layout cues. Modern cell phones may display a half dozen lines of text, each containing about 25 characters, apparently enough to support reading and text entry for informal text messaging.

4.6 NAVIGATING THROUGH TEXT WITH MAGNIFIERS

Some people with mild forms of vision impairment read without magnifiers and use ordinary eye movements for reading. We begin by discussing low-vision reading without magnifiers.

Using Large Print or Short Viewing Distance to Magnify

In section 3.2, we pointed out that newsprint, viewed from 40 cm, is close to the CPS for people with normal acuity. Typical newsprint has an x-height of about 1.45 mm, equivalent to 1 M-unit (DeMarco & Massof, 1997). People with reduced acuity would need some form of magnification to read the newspaper effectively at a 40 cm viewing distance. There are three general forms of magnification for reading—enlarge the print size on the page, bring the eye closer to the page, or use a magnifier.

Word processors and some copiers can be used to enlarge print size. Some published materials are available in "large print" formats. The recommended print size for large-print materials is 16 to 18 points (Arditi, 1999b). Helvetica and Tahoma are two fonts commonly used in large print materials. Table 4.3 shows our measurements of the lowercase x-height and uppercase X-height for large-print in these two fonts. The corresponding sizes in M-units are also shown. These heights range from about 3 mm to 4mm, roughly 2 to 3 M-units. These print sizes are roughly two to three times the size of typical newsprint. People needing more than three-times magnification for newsprint would find large print materials difficult to read without additional magnification.

If large print is not available (frequently the case) or if it provides insufficient magnification, a low-tech method of magnification is to bring the eye closer to the page, thereby reducing viewing distance. Every low-vision person makes use of this technique from time to time. Since angular character size is inversely related to viewing distance, reduction of the viewing distance by a factor of N accomplishes N-fold magnification. For instance, viewing the newspaper from 20 cm rather than 40 cm (a factor of two) is equivalent to magnifying print size by a factor of two.

Here is a simple formula that shows how physical print size and viewing distance combine to determine magnification W:

$$W = 40S/D,$$

where S is print size in M-units, D is viewing distance in cm, and W is magnification relative to typical newsprint of size 1 M-unit viewed from 40 cm. For instance, if the print size S is 2.8 M-units (such as Tahoma x-height at 18 pt), and the viewing distance D is 15 cm, then the magnification W is 7.47 (i.e., 40 × 2.8/15). In other words, the angular character size (and corresponding retinal-image size) is 7.47 times larger than typical newsprint viewed from 40 cm (Do not confuse this definition of magnification with the convention for defining the magnification of an optical magnifier described in footnote 18 of this chapter).

TABLE 4.3
Examples of Sizes of Large Print*

Font/ Point Size	X-height (mm)	M-Units	x-height (mm)	M-units
Helvetica (16 pt)	3.90	2.70	2.80	1.90
Helvetica (18 pt)	4.10	2.80	3.00	2.10
Tahoma (16 pt)	3.90	2.70	2.90	2.00
Tahoma (18 pt)	4.10	2.80	3.10	2.10
Newsprint[+] (typical size)	—	—	1.45	1.00

*x-height in mm is converted to M-units by dividing by 1.454. For other print-size conversion formulas, see Appendix A.

[+]DeMarco and Massof (1997) conducted a survey of print sizes in U.S. newspapers. They found that the median size of print ranged from 0.78 M-units to 1.21 M-units, depending on the section of the newspaper. In this table, we have listed the "typical size" of newsprint as 1.0 M-units.

From the above analysis, it would seem a simple matter to deal with all magnification needs by some combination of large-print size and short viewing distance. Problems quickly emerge with both of these manipulations. First, the vast majority of published material is not available in large print. Even when large-print publication is an option, there are practical limitations on the sizes of pages and books. For this reason, large print rarely exceeds 20 pt. Second, when the reader adopts a shorter viewing distance, it is necessary to re-focus for the nearer distance. Young people who have a wide range of accommodation, or people who are myopic (short-sighted), may be able to focus at distances of 10 or 20 cm. Most people, especially older people with presbyopia (the absence of accommodation, encountered by almost everyone over 50), will require an external plus lens to focus print at short viewing distances. Plus lenses used in this way are term *magnifiers*.

Reading With Magnifiers

The issue of page navigation comes to the fore whenever magnifiers are required for reading. As we discussed in section 4.4, enlargement of the characters in a local region of text by a magnifier brings about the need to move the magnifier through the text. Use of a magnifier necessitates combining eye movements with some other form of movement. For reviews of properties and principles for prescribing low-vision magnifiers, see Bailey, Bullimore, Greer, and Mattingly (1994) and Sloan (1977). Spectacle-mounted optical magnifiers require head movements and/or manual panning of the text through the field of view (cf. A. R. Bowers, 2000). A hand-held magnifier or stand magnifier requires manual scanning. As discussed in section 4.4, CCTV magnifiers or computer screen magnifiers usually require manual movements of a CCTV platform or mouse.

As we discussed in section 2.1, ordinary eye movements impose a ceiling on normal reading speed, a ceiling that can be lifted with the use of RSVP. It is reasonable to anticipate that the added page-navigation demands of a magnifier would lower the ceiling on reading speed below values for ordinary reading. Several studies have shown that normally sighted participants slow down when reading with optical or electronic magnifiers—by 35% to 40% for hand-held or spectacle-mounted magnifiers (A. R. Bowers, 2000), and by a little over 40% when reading with a stand magnifier (E. M. Fine et al., 1996). Although stand magnifiers reportedly cause a greater reduction in speed than hand-held or spectacle-mounted magnifiers (J. M. Cohen & Waiss, 1991a; Mancil & Nowakowski, 1986), the difference disappears if magnifiers are equated in field size (J. M. Cohen & Waiss, 1991b). In R14 (1996, Fig. 5), we found that a group of normally sighted participants had a mean CCTV reading speed of 140 wpm, roughly 50% below their ordinary reading speeds. Summarizing across these studies, it appears that the use of a magnifier reduces the upperbound on normal reading speed by about 30% to 50%. But what is the effect on low-vision reading speed?

Given ample training with a magnifier, how much do the demands of page navigation limit a low-vision person's reading speed? In R14 (1996), we studied the impact on reading speed of page navigation with a CCTV magnifier. We added two optical encoders to the movable platform of a CCTV magnifier, one measuring the platform's front-to-back position and one measuring its side-to-side position. The accuracy was 0.5 mm. A computer sampled and recorded these measurements at every 100 msec. The resulting stream of data produced waveforms analogous to eye-movement recordings. For an example, see Figure 1 in R14.

Following a period of practice with the CCTV apparatus, normally sighted and low-vision participants read passages composed of 13 lines of text. For each participant and passage, we split the total reading time into two components: the time taken in moving from the beginning of an individual line to the end of the line ("forward reading time"), and the time to return from the end of one line to the beginning of the next line ("retrace time"). We reasoned that reading with a magnifier involves two main processes, visual decoding of letters and words, and the visual-motor demands of moving the magnifier over the text. Only the latter is operative during the retrace time.

Our normally sighted participants had best reading speeds of about 140 wpm (group mean) using CCTV, at least a factor of two below their ordinary reading speeds. For these participants, the forward and retrace times were about equal, and were probably both limited by the demands of manual page navigation. Because half of the total time was required for retracing, overall reading speed was reduced by a factor of two due to retrace time. By comparison the proportion of total reading time taken up by retracing in ordinary eye-movement reading is only about 5% (A. R. Bowers, 2000).

For the low-vision participants in our study, the forward reading times were usually substantially longer than the retrace times. This difference implies that the low vision participants were limited by visual decoding of text, unlike the normal participants who were limited by the demands of page navigation. Moreover, because the retrace times occupied a smaller proportion of total reading time for the low-vision participants, retrace time had less impact on their overall reading speeds. For instance, the retrace time for low-vision participant C occupied only 17% of the total reading time, representing a small impact on overall reading speed.

A. R. Bowers (2000) assessed the impact of retrace time on reading speed for normally sighted participants who read with hand-held and spectacle-mounted magnifiers. Unlike our experience with normally sighted participants and CCTV, Bowers's participants made faster retrace movements than forward movements with their magnifiers; only about 20% of the total reading time was devoted to retracing. In the case of CCTV reading, it is likely that normally sighted participants are limited in both the forward reading and retrace directions by manipulation of the CCTV platform. In the case of optical magnifiers, it may be possible to make rapid (saccade-like) head or hand movements in retracing from the end of one line to the beginning of the next line.

We concluded from our CCTV study (R14, 1996) that for most participants with low vision, visual decoding imposes the primary constraint on reading speed, with the demands of page navigation having a relatively modest impact. Although the encouraging message is that a user's manipulation of a well designed and properly prescribed magnifier should not limit reading speed, a possible implication is that technology for reducing page-navigation demands might have less beneficial effect in low vision than in normal vision. This would be the case if visual decoding imposes the fundamental limitation on reading speed.

We investigated this implication in a study comparing performance with four types of electronic text presentation (R17, 1998). The four methods varied in their page-navigation demands. Two methods involved manual navigation—CCTV, and use of a computer mouse for page navigation as used in computer screen-magnification software. The third method, drifting text, involved no manual navigation, but did involve both smooth-pursuit and saccadic eye movements. The fourth method, RSVP, involved no manual navigation, and minimal eye movements. The visual-display parameters and task (reading a passage of text for understanding) were the same for all four text-presentation methods. There were seven participants with normal vision and 12 with low vision (7 with central-field loss and 5 with residual central vision). All the participants received practice with the four methods prior to measurement of reading performance.

We used reading speed for CCTV as a benchmark for comparison with the other methods. Neither the normally sighted nor the low-vision participants showed a difference in reading speed between the two manual methods, CCTV and Mouse. As expected, the normal participants read faster with the two methods not requiring manual page navigation—85% faster with drifting text, and 169% faster with RSVP. This pattern is consistent with the expectation that a reduced page-navigation demand should result in faster reading. The results reviewed above from R14 (1996) imply that people with low vision might show less benefit from reduced page-navigation demand. This is what we found. The low-vision participants with residual central vision did read somewhat faster with drifting text than with CCTV (43%) and with RSVP (38%). The participants with central-field loss showed no statistically significant difference in reading speed across the four text-presentation conditions.

We mention one additional finding from our study (R17, 1998). We asked our participants to rank order their preferences for the four types of text presentation. The order of mean preferences was the same for normal and low vision—drift (best), CCTV, mouse, and RSVP (worst). Dislike of RSVP, despite the speed advantage for normals and some participants with low vision, is consistent with findings cited earlier in this chapter from participants with normal vision regarding complaints about RSVP as a method for displaying text.

Three other studies have also shown that seemingly major differences in the nature of page navigation do not result in major differences in low-vision reading speed. Bowers et al. (2004) assessed the impact of four types of text presentation on the reading speeds of people with low vision, most of them with central-field loss. The four types were RSVP, horizontal scroll (the same as our drifting text), vertical scroll (words drifted vertically upward like movie credits), and page format (3 lines of text visible at a time, each containing 2 or 3 words, with 2 or 3 such pages required to display a 56-character sentence).[27] They found no significant differences in maximum oral reading speed across the four types of text presentation for the low-vision participants.

Ortiz, Chung, Legge, and Jobling (1999) compared low-vision reading speed with CCTV and with a head-mounted video magnifier known as LVES (Low Vision Enhancement System).[28] For LVES reading, the participants used a chin rest to maintain their heads at a distance of 25 cm from test passages mounted on an easel. Page navigation was accomplished with a combination of head movements to change the direction of the camera, and eye movements to scan the video image. Despite complaints from the participants about the uncomfortable LVES apparatus on their heads, their maximum reading speeds for LVES and CCTV were not significantly different. Subsequently, Culham, Chabra, and Rubin (2004) evaluated four types of headmounted video magnifiers, and compared them with optical low-vision aids. They tested groups of participants with early-onset and age-related forms of macular degeneration. These participants used the magnifiers to perform several clinical tests and everyday tasks including reading. Reading speeds were measured for three print sizes. For most of the reading tests, the participants read faster with an optical low-vision aid than with any of the head-mounted video magnifiers. There was one exception; the participants with early-onset macular degeneration read the tiniest print (smaller than typical newsprint) faster with one of the headmounted magnifiers (Flipperport) than with their optical low-vision aid. From these two studies, it appears that the current generation of head-mounted video magnifiers are rarely better for everyday reading than existing optical or CCTV magnifiers.

Finally, we mention two variants of RSVP that have demonstrated some benefits for low-vision reading. Aquilante, Yager, Morris, and Khmelnitsky (2001) modified RSVP so that presentation times varied with word length rather than being equal for all word lengths. With this modification, participants with AMD read 33% faster. Another variant of RSVP is the elicited sequential presentation (ESP) method developed by Arditi (1999a). In this procedure, the participant triggers the presentation of each word with a key press. Like RSVP, the words appear sequen-

---

[27]In this study, information was obtained on the position of the preferred retinal locus (PRL) of subjects with central scotomas. A major purpose of the study was to ask whether position of the PRL interacts with the type of text presentation in determining reading speed. For instance, it was hypothesized that horizontally drifting text would be a better reading stimulus than vertically drifting text for people with PRLs above or below a scotoma, and that vertically drifting text would be a better stimulus when the PRL is left or right of a scotoma. Contrary to expectation, there was no statistically significant interaction of PRL location and type of text presentation.

[28]LVES consists of a headset (weighing 2.6 pounds) containing battery-powered, black and white, miniature video displays for each eye. It is equipped with two video cameras having fixed-focus unit magnification optics for orientation and a center-mounted camera with variable focus and variable magnification optics for performing near, intermediate and distance tasks. LVES was the first example of a portable, head-mounted video magnifier for low vision. Although it is no longer sold commercially, it has inspired the development of newer, lightweight, head-mounted electronic magnifiers.

tially at the same place on a display screen, but unlike RSVP, ESP gives people control over the delivery of words, allowing them time to position their eyes or extra time to deal with difficult words. We have argued that reduction of the size of the visual span in low vision should mean that longer words take more time to recognize (R16, 1997). Consistent with this expectation, Arditi (1999a) did find a significant correlation between ESP display time and word length, although the relationship was not very strong. Arditi (1999a) found that all of his 15 low-vision participants read faster with ESP than with RSVP, the difference ranging from 4% to 122% with a group mean of 47%. All of the participants in this study were experienced CCTV users. Although they read faster with ESP than with RSVP, the ESP reading speeds did not differ significantly from CCTV reading speeds. Even if ESP is considered to be an improved version of RSVP, as demonstrated by Arditi (1999a), we do not yet have evidence that it yields faster reading than CCTV.

We summarize the discussion in this subsection with five conclusions about navigation through text with magnifiers:

- The extra page-navigation demand of magnifiers reduces the upper bound on reading speed below values for ordinary eye-movement reading.
- The large differences in reading speed associated with different types of page navigation for normal vision are compressed or absent in low-vision reading.
- After people with low vision have learned to use magnifiers, the extra page-navigation demands will not usually have a large impact on reading speed.
- Automated methods for displaying text, such as RSVP or drift, are not likely to yield large benefits in low-vision reading speed.
- There do not appear to be inherent differences in reading speeds for the three major types of optical magnifiers—stand, hand-held and spectacle-mounted—as long as the visual properties (especially field size and magnification) are equivalent.

Hypertext Retrieval With Low Vision and the Local–Global Problem

Stereotypical reading involves the sequential processing of continuous text, word by word, and line by line. The reader knows where to look next because the words are arranged in a predictable layout from left to right and top to bottom on the page. This predictable arrangement of text is particularly helpful for magnifier users because there is little uncertainty about where to move the magnifier next. Reading, however, need not be restricted to sequential processing of text, and frequently involves skimming for gist, or search for specific terms or factual items of information.

Hypertext reading typically involves search and retrieval and is an essential aspect of Web browsing. It typically involves nonsequential processing of text, emphasizing the need for rapid skipping from place to place in a complex page layout composed of a mix of text, graphics and hyperlinks.

The "local–global" problem, inherent in most magnifiers, makes Web browsing, and other types of nonsequential reading, especially difficult for people with low vision. This problem is illustrated in Figure 4.11. The left panel shows a typical complex Web site, the Science Magazine home page. If a person with low vision uses screen-magnification software to inspect the display, a local region is enlarged, right panel, but the rest of the image disappears from view. Given only this magnified local view, the global arrangement of the Web page into columns is hard to understand, and it would be a guess where to look for the Search and Browse features or the date of the current issue. If an error message or other informative cue, such as the location of a scroll bar, appears outside the magnified local view, it may go entirely unnoticed. To understand the Web page, the user has the burden of moving the magnified window over the display, and in piecing the sequence of local views together to infer the global layout. This need for page navigation, and the possibility that informative cues or alerts appear off screen are major impediments to reading Web pages or other complex layouts with magnifiers. The local–global problem also afflicts normally sighted people who use small displays on a cell phone or PDA for Web browsing.

Almost all of the psychophysical research on text reading has used the stereotypical sequential reading task. An exception is our study of hypertext search and retrieval (R19, 2002). The goal of this study was to determine if people with low vision are at a more severe speed disadvantage in hypertext retrieval than in conventional prose reading. We created two text-only Web sites running on a local server, one dealing with juggling and the other with low vision. The two Web sites had identical hierarchical structures, and consisted of 54 pages (termed "nodes" in the study) each occupying a single screen, and 53 hyperlinks (see R19, 2002, Fig. 1). The participant's task was to find the answers to a series of factual questions by searching the Web site. Low-vision participants viewed the Web site with a screen

Regular screen view

Screen magnifier view

Figure 4.11   Illustration of the local–global problem of magnifiers. The left panel shows the internet home page for Science Magazine, http://www.sciencemag.org/, and the right panel a magnified portion of the same home page occupying the same display area.

magnifier containing a magnified local view 12 characters in width. They used the computer mouse to move the magnified window over the Web page. Sighted participants viewed the normal screen display. The server recorded the sequence of pages traversed and the time taken on each page.

In Experiment 1 of R19 (2002), we compared the reading times of groups of normal and low vision participants in two tasks: conventional prose reading (sequential reading), and hypertext retrieval (nonsequential reading). In this experiment, the hyperlinks were all located in a predictable left-justified column at the edge of the screen. For the prose reading task, the average reading speed for the low-vision participants was 30% of the average reading speed for the normally sighted participants, (61 and 201 wpm respectively) that is, the low-vision participants took about 3.3 times longer to read the test passages than the normal participants. Surprisingly, the time ratio was almost the same for hypertext retrieval; the low-vision participants averaged 3.2 times longer than the normally sighted participants to find the answers to the test questions. This result showed that hypertext retrieval per se, involving moving from page to page via a series of hyperlinks, does not impose an extra time cost on low-vision readers.

Experiment 2 was designed to investigate the effect of reducing the predictability of the layout of hyperlinks on low-vision performance. Low-vision participants performed the same type of hypertext retrieval task with two arrangements of hyperlinks: links in one left-justified column as in Experiment 1, and links distributed unpredictably across left, center and right positions on the screen (R19, 2002, Fig. 4). The average search time was substantially longer for the unpredictable arrangement of hyperlinks (106 sec) compared with the predictable arrangement (65 sec). The extra time was presumably used by the low-vision participants in searching with the magnified window for the relevant hyperlinks.

We also tested a third condition, intended to offset the adverse effect of the unpredictable arrangement of links. The low-vision participants were given a split-screen display (R19, 2002, Fig. 5). The lower half of the screen showed a magnified view of the text at the cursor location, and the upper half showed an unmagnified view of the screen. Although the low-vision participants could not read the text in the unmagnified upper view, they could discern the global page layout of the text and the locations of hyperlinks. The hyperlinks contrasted in color with the text and were easy for low-vision participants to distinguish from the rest of the text. The participants used the unmagnified global view to spot the hyperlinks and steer the cursor to them, and then they used the magnified window on the lower half of the screen to read the hyperlinks. With this split-screen arrangement the low-vision participants achieved search times for the case of unpredictable link locations that were virtually identical to the case of the predictable link arrangement. Our interpretation was that our low-vision participants could effectively use the two distinct types of information provided by the split-screen display: the global page layout on the upper screen, and the magnified text on the lower screen. In this case at least, the local–global problem associated with the use of magnifiers was solved by presenting local and global information in two nonoverlapping displays.

The availability of split-screen views is a feature of versatile screen-magnification software for computers.

To summarize, the results of R19 (2002) showed that low-vision hypertext search is much slower for unpredictable layouts of hyperlinks compared with a predictable columnar arrangement. Fortunately, however, access to global layout through a split-screen configuration results in a reduction in search time that offsets the extra time associated with unpredictable layouts.

4.7 COLOR AND LUMINANCE EFFECTS IN NORMAL AND LOW VISION

People enjoy using color to spruce up text. Does color affect text legibility? People buy reading lamps to make text easier to read. What is the effect of light level on legibility? We address these questions in this section.

One challenge in understanding the literature on these topics is the bewildering assortment of units for measuring color and light intensity. Box 4.3 briefly comments on units of light intensity often used in studies of reading.

Color

Does the color of text affect reading performance? In R4 (1986), we reported on the effects of text color on reading speed for four normally sighted participants, two participants with color blindness (both dichromats), and 25 low-vision participants. Most of the measurements were conducted with the drifting text method.[29] We compared performance under four luminance-matched (6 cd/m^2) color conditions. Colored text (colored letters on a black background or vice versa) was produced by placing Wratten broad-band color filters in front of a grayscale video screen showing the text. The four filters were blue (λ_{max} = 430 nm), green (λ_{max} = 550 nm), red (λ_{max} = 650 nm), and gray.

For the normally sighted participants, there was no effect of wavelength on reading speed for highly magnified letters (6°) at photopic levels[30] (R4, 1986; Fig. 2). For small letters very close to the acuity limit, two of the four normal participants showed depressed reading speed in the blue, probably because their acuity was poorer for short wavelength than for long wavelengths (Pokorny, Graham, &

---

[29]Limited control experiments on two of the normally sighted subjects showed no differences in wavelength effects for measurements with stationary text.

[30]Under scotopic conditions, our normally sighted subjects were obliged to use peripheral vision because the foveal cones were unresponsive. (One of our most dedicated subjects was disturbed by the scotopic viewing and accused us of "playing tricks" on him because the text disappeared whenever he tried to look at it. We explained that it was normal for central vision to become inoperative under very low illumination and that he would need to use his peripheral vision to read). Despite the need for peripheral vision, our normally sighted subjects could read at all wavelengths at a scotopic level of 0.006 cd/m^2. Under these conditions, reading speeds were higher in blue than the red, because of the greater scotopic spectral sensitivity of the eye at short wavelengths.

Lanson, 1968). One of our two dichromats[31]—a protanope with depressed spectral sensitivity at long wavelengths—exhibited reduced reading speed in the red. Our other dichromat, a deuteranope, showed no wavelength-specific reading effects. We concluded that for luminance-matched text, there is no important effect of color (wavelength) on photopic reading speed for normally sighted participants, except for letters very close to the acuity limit.

In a subsequent study, Pastoor (1990) examined participants preferences for many different combinations of text and background colors, and for two text polarities (bright letters on dark backgrounds and dark letters on bright backgrounds.) He also measured reading times for a subset of the color combinations (Experiment 2). In agreement with R4 (1986), Pastoor found no influence of the color combination or polarity condition on reading time.

In R4 (1986), we also examined color effects on reading speed for four groups of low-vision participants. All were tested with large 6° characters at photopic luminance. One group consisted of seven participants with light scatter and other problems of the ocular media. We predicted that they would show selective deficits for short wavelength blue light because of greater scatter or absorption in the eye. Only one of the seven participants (participant A) confirmed the prediction; the others showed no color effects on reading. Follow-up experiments with participant A showed that his reading deficit for blue text was related to increased light absorption at short wavelengths by the optics of his eye.

A second group of six low-vision participants had degenerative photoreceptor disorders (four of them with retinitis pigmentosa). For four of the six participants, reading speed was fastest in the blue and declined at longer wavelengths. Additional groups of five participants with central field loss and six participants with peripheral field loss exhibited remarkably little effect of wavelength on reading speed.

Taken as a whole, the results of R4 (1986) indicate that relatively few people with normal or low vision exhibit effects of color (wavelength) on reading speed when luminance levels are matched. In the exceptional cases in which wavelength effects are present, reading speed is more likely to be depressed for short wavelength blue or long wavelength red text, than for medium wavelength green or broadband black-on-white text. The implication is that the best colors for the design of text displays are black-and-white or black-and-green.

Text color usually refers to black letters on a colored background or colored letters on a black background. In both cases, the characters and background differ in luminance and the text has high luminance contrast. As reviewed in detail in section 3.3, rapid reading is possible if the luminance contrast of text exceeds about 10%. In R11 (1990), we asked whether text can be read if the letters and background have equal luminance (luminance contrast is zero), but background and letters differ in color, such as red letters on a green background. The answer is yes.

---

[31]Dichromats are individuals who have only two of the three cone pigments for color vision. The three types of dichromats are characterized by the lack of cone pigments peaking at short, medium and long wavelengths and are termed tritanopes, deuteranopes and protanopes respectively.

For normally sighted participants, reading speeds for equiluminant text with high chromatic contrast can match those for text with high luminance contrast (R11, 1990; Knoblauch et al., 1991; Travis, Bowles, Seton, & Peppe, 1990). Keep in mind, however, that a substantial minority of people have inherited color defects that make some color differences, particularly red-green differences, hard to see. For this reason, designing text displays with pure color contrast can be risky.

In R11 (1990), we also reported on low-vision reading with color contrast. For all ten participants with low vision, reading speeds with equiluminant color contrast were always lower than reading speeds for text with high luminance contrast. Although bright colors can sometimes provide helpful cues for people with low vision in other circumstances, the results from R4 and R11 provide no evidence that color coding enhances reading speed. Based on our current understanding, maximizing luminance contrast is the best way to enhance text legibility for low vision.

Here are three guidelines for selecting contrasting colors to maximize display accessibility for people with congenital color defects and color-vision problems from low vision (Arditi, 2005). These guidelines apply to any display in which pairs of colors are used, including text:

- Maximize the luminance contrast between the two colors.
- Choose the light color of the pair to be white or predominantly from the middle of the visible band of wavelengths (green or yellow). Choose the dark color of the pair to be black or predominantly from the ends of the visible spectrum (red, blue, or purple.) The logic is that eye diseases or congenital color defects are more likely to depress spectral sensitivity at the ends of the visible spectrum, so it is best to use these for the darker color component of the contrasting pair.
- Choose the hues to be far apart in color space, that is, use high chromatic contrast.

Colored Overlays

The educational psychologist Helen Irlen (1991) developed a widely publicized method for using colored overlays or tinted lenses to relieve the symptoms of visual discomfort and print distortions in reading. The visual problem addressed by this treatment has been termed scotopic sensitivity syndrome, or Irlen Syndrome. The existence of this syndrome and the benefits of colored overlays for reading has been controversial within the scientific community. The most comprehensive series of studies has been conducted by Arnold Wilkins and colleagues. For a brief review, see Wilkins (2002).

Wilkins (1994) described a test composed of "intuitive overlays" to assess the preferences of school children for relieving symptoms of visual discomfort in reading. The test consists of 9 colored overlays and a 10th gray overlay. These ten overlays, sometimes used singly and sometimes in overlapping pairs were chosen to systematically sample the Commission Internationale de l'Eclairage Uniform

Chromaticity Scale (CIE-UES) color diagram. The colored overlay is placed on the printed page for reading, giving standard text the appearance of black print on a colored background. In a study of 426 young children (mean age 7 years, 6 months), about 60% preferred reading with a colored overlay to no overlay, and after several months about 31% of the original sample were still using an overlay (Wilkins, Lewis, Smith, Rowland, & Tweedie, 2001). Children who used the overlays tended to be those who reported symptoms of discomfort in reading including movement of letters, lack of clarity, or glare from the page. The preferred color of overlays varied widely across the children and did not cluster systematically in any confined region of the color space.

Wilkins et al. (2001) reported on the impact of colored overlays on children's reading speed in three studies. Reading speed was measured with a test composed of 15 familiar words, repeated in different random orders on 10 lines. Children were timed as they read these passages aloud with and without colored overlays.[32] In study 1, 78 out of 89 children (mean age 9 years, 4 months) selected a colored overlay for reading, and exhibited an average increase in reading speed of 11%. In Study 2, 83% of 378 school children (age range from 8 years 2 months to 12 years 1 month) chose a colored overlay. Across the group, the mean reading speed with the preferred overlay was 116.8 wpm compared with 112.7 wpm with no overlay, a statistically significant but modest improvement of 3.6%. In Study 3, 426 children were tested, 273 of whom reported a benefit from overlays. For this group, the mean reading speed with the chosen overlay was 74.43 wpm, and without overlay 71.73 wpm, representing a mean improvement of 3.7% in reading speed. The authors report that 5% of the children read more than 25% faster with their overlays, but it is unclear how this figure compares with the distribution of simple test-retest differences to be expected for young children on this test of reading speed. Across the three studies, the authors point out that children who were consistent in selecting the same overlay across repeated testing, or those who persisted in using overlays over time, tended to show somewhat greater improvements in reading speed with overlays. In another study, Robinson and Foreman (1999) tracked the reading performance of 113 children with reading difficulties characteristic of those associated with scotopic sensitivity/Irlen syndrome, and also 35 controls, for 20 months. The children with reading difficulties were randomly assigned to treatment with optimally colored overlays, placebo overlays (with similar but nonoptimal color) and blue overlays. The controls used no overlays. There was no significant difference in the improvement in reading speed across the four groups, although the treatment groups did show greater improvements in reading accuracy and comprehension.

The physiological basis of improvements in reading comfort or speed due to colored overlays remains unclear. The effects are not related to the well-studied inherited color-vision defects. Proposed mechanisms that appear unlikely include

---

[32]Wilkins and colleagues argue against placebo effects, order effects, practice effects and examiner effects as explanations for improvements due to colored overlays (Wilkins, 2002: Wilkins et al., 2001).

binocular abnormalities, problems with ocular accommodation, and retinal deficits (Wilkins et al., 2001). The wide individual variations in optimal overlay color
is inconsistent with any straightforward version of the magnocellular deficit theory (Wilkins et al., 2001). One plausible explanation links discomfort in reading to
visual stress in migraine and epilepsy (Wilkins, 1995). The link is forged through
the known tendency of striped patterns such as square-wave gratings), to induce
photosensitive seizures, and the similarity of these patterns to lines of text. Wilkins
et al. speculated that "... comfortable colors reduce strong excitation in hyperex-
citable regions (of visual cortex), reducing an inappropriate spread of excitation
..." (p. 62).

The cited studies indicate that the effect of colored overlays on children's reading speed is quite weak, although potentially stronger for a small proportion of
children who benefit most. The relatively weak effect on reading speed suggests
that colored overlays do not directly influence the visual decoding of words and
letters in reading, but may have some collateral influence on reading comfort. It is
not known if colored overlays have similar effects on reading speed and reading
comfort for adults.

Luminance

The National Research Council (2002) recommended a luminance of 160 cd/m^2
(and not less than 80 cd/m^2) for acuity charts. IESNA, The Illuminating Engineering Society of North America, recommended a luminance for reading in the range
from 75 to 125 cd/m^2 (IESNA, 2000).[33]

Reading light levels may need to be higher for older people because of smaller
pupil size, and greater light absorption by the optics of the eye. According to a frequently cited report, the average 60-year-old eye transmits about one third the
amount of light of the average 20-year-old eye (Weale, 1963, p. 168). Of course,
this rule of thumb is subject to wide individual differences, and effects of stimulus
variables such as wavelength (shorter wavelengths will be absorbed more by the
older eye.) Nevertheless, if we specify 100 cd/m^2 as a target light level for reading
by young normal participants, it is prudent to recommend about 300 cd/m^2 for old
normal readers.

How do these recommended values compare with light levels in real-world
reading environments? Charness and Dijkstra (1999) made field measurements of
typical lighting levels for reading in homes, business offices and public places in
Tallahassee, Florida. They found that home lighting was often lower than the recommended level of about 100 cd/m^2, averaging 37 cd/m^2 for daytime measurements and 15 cd/m^2 at night. Office measurements were better (mean 143 cd/m^2),

---

[33]The range of luminances cited here are inferred from IESNA's recommendation of 300 lx for Category D tasks and 500 lx for category E tasks (IESNA, 2000, p. 10.13). The majority of reading situations
fall within these two categories. The conversion from illuminance, measured in lx, to luminance, measured in cd/m^2 used the conversion formula given in Box 4.3.

Box 4.3
Page Luminance and Illuminance

Lighting engineers refer to the illumination of a surface, such as a page of text, by the light flux reaching the surface. *Illuminance* is the luminous flux per unit area incident on a surface. In metric units, it is measured in lux (lx). (To convert illuminance in foot-candles to lux, multiply by 10.76.)

Vision researchers usually refer to the luminous intensity reaching the eye from a light source or the reflected light from a surface such as a page of text. *Luminance* is the luminous intensity emitted per unit area from the light source or reflective surface. In metric units, it is measured in candelas per square meter, cd/m^2. (To convert luminance in foot-lamberts to luminance in candelas per square meter, multiply by 3.43.) To convert the illuminance of a surface to the luminance of the surface seen by the eye, we need to know the reflectance of the surface. Reflectance is the ratio of the reflected luminous flux to the incident luminous flux, and is a number between 0 and 1.0 (or 0 and 100%). White paper has high reflectance, gray paper medium reflectance and black paper or black ink low reflectance. According to IESNA (2000), the reflectance of common printed materials typically ranges between 30% and 70%.

The formula for converting illuminance I in lx to luminance L in cd/m^2 for a surface with reflectance R is:

$$L = IR/\pi$$

For a white surface with a high reflectance of 0.785, this formula reduces to this simple conversion:

$$L = I/4.$$

In other words, for an approximate conversion from surface illuminance in lux to luminance in candelas per square meter, for a surface with high reflectance, such as a white page of paper, divide the illuminance value by 4. For instance, for a page illuminated by 400 lx, the luminance of the white page is approximately 100 cd/m^2. For a more detailed discussion, see Wyszecki and Stiles (1982, p. 262).

---

but 70% of the measurements in public spaces (taken at reception desks and in waiting rooms) were below the recommended level of 100 cd/m^2.

How sensitive is reading speed to luminance? We conducted limited studies of this question, mostly at low photopic levels and very large print sizes. Our studies revealed that reading speed is at most weakly dependent on photopic luminance for young, normally sighted participants (R1, 1985; R4, 1986). Bullimore and Bailey (1995) measured reading speed as a function of print size for six older participants (range 62 to 77 years) with normal vision. The maximum reading speeds of about 160 wpm were very similar at the two tested luminances of 1 and 100 cd/m^2, but the CPS was smaller at the higher luminance. This finding illustrates that reading speed and CPS may have a different dependence on luminance level (and potentially other stimulus variables).

Some people with low vision read faster, or have smaller CPSs, at higher levels of illumination, although there are wide individual variations in the optimal levels (A. R. Bowers, Meek, & Stewart, 2001; Eldred, 1992; LaGrow, 1986). A. R. Bowers et al. (2001) measured reading speed as a function of print size (MNREAD method) for 20 patients with AMD. Testing was conducted at six levels of illumination ranging from 50 to 5,000 Lx. See Box 4.3 for a brief discussion of conversion from illuminance in lx to luminance in cd/m^2. Both reading acuity and CPS improved by a factor of two over this range, while average reading speed increased by about 40%. Most of the improvement occurred for illumination of 2,000 Lx or less, putting a practical bound on the desirable level of illumination for this group. This level of 2,000 Lx compares with typical values of 50 Lx for page illumination in the home, and 500 Lx in the eye clinic (A. R. Bowers et al., 2001). These findings confirm that AMD patients frequently benefit from supplementary reading lights.

Why should AMD patients have better acuity and read faster at higher light levels? Bullimore and Bailey (1995) have shown that the size and morphology of central scotomas in AMD patients depend on light level (see their Fig. 8). For a decrease from 100 to 1.0 cd/m2, there was an 80% increase in median scotoma size for their sample of AMD patients. In some cases, changes in morphology included the emergence of a foveal island of vision at the higher light levels. Lei and Schuchard (1997) showed that such patients may use different preferred retinal loci (PRLs) for low and high luminance, with the high-luminance PRL tending to be closer to the fovea. These studies indicate that the improvement in AMD reading at high light levels is probably due to changes in the characteristics of central scotomas.

Finally, we list three interacting factors to optimize lighting for low-vision reading. First, high luminance contrast of the text is often necessary. Second, high overall luminance is sometimes necessary. Third, for some people, contrast reversal (bright letters on a dark background) is preferable (see section 3.3). Sometimes, efforts to maximize lighting can have adverse effects. For instance, strong room lighting intended to optimize luminance for page reading may produce a veiling glare that dilutes luminance contrast on a computer display. Bright windows or task lights in the field of view, intended to increase lighting on a page of text, can cause glare effects that dilute retinal-image contrast.

4.8 GUIDELINES FOR LEGIBLE TEXT

In chapters 3 and 4, we have discussed many stimulus attributes relevant to displaying text. We have focused on reading speed as the primary measure of legibility. In this section, we summarize the most important empirical findings for easy reference, and include pointers to the chapter sections containing the corresponding discussion

Critical Values for Reading Speed

Several of the most important stimulus dimensions have critical values (or thresholds) necessary for maximum reading speed.

TABLE 4.4
Critical Points

	Normal Vision	Low Vision
Character Size (Sec. 3.2)	Optimal range is $0.2°$ < character size < $2°$. The small end of this range is termed *critical print size*.	Critical print sizes vary widely, but generally increase as the size of acuity letters increase.
Contrast (Sec. 3.3)	5-10 % in the range of optimal character size. Critical contrasts are higher for very tiny or very large characters.	Scale up the critical contrast for reading by a factor corresponding to the loss of contrast sensitivity.
Spatial-Frequency Bandwidth (Sec. 4.3)	2 cycles per character. For characters subtending A degrees, this corresponds to 2/A cycles per degree.	Not studied.
Sample Density (Sec. 4.3)	20×20 samples per character is a safe upper bound. Critical values depend on character size ranging from 4×4 for tiny characters to 20×20 for very large characters.	20×20 samples per character is desirable.
Window Size (Sec. 4.4)	5 characters, if no page navigation is necessary. 12 characters if page navigation is required.	5 characters if no page navigation is necessary and 8 to 12 characters if page navigation is required.

Font Characteristics

A few general principles may help in the selection of fonts for normal and low vision.

Character Shapes (Section 4.2).   The featural requirements for legible fonts are very loose. Any set of characters should do, as long as they contain a sufficient number of simple, easily detectable features in distinct spatial configurations.

Fixed Width Versus Proportionally Spaced (Section 4.2).   For normal vision, proportionally spaced fonts yield slightly faster reading speed for character sizes larger than the CPS. Fixed-width fonts are faster when the character size is smaller than the CPS. Low-vision reading is faster with fixed-width fonts, but the extra space used by these fonts imposes practical limitations on their use in books or other printed material.

   Character Spacing (Section 4.2).   The more ample spacing around narrow letters in a fixed width font compared with a proportionally spaced font appears to facilitate reading in low vision and also in normal vision when reading near the acuity limit. But adding extra spacing to standard fonts does not have practical advantages for reading in normal or low vision.

   Anti-Aliasing, Grayscale Smoothing (Section 4.3).   This has little or no benefit for reading speed for standard or small print size, but may be beneficial for large characters used by people with low vision.

Color and Luminance

Attention to color and luminance properties of text can enhance legibility for people with normal and low vision.

TABLE 4.5

Color and Luminance Properties of Text

	Normal Vision	*Low Vision*
Text Color (Sec. 4.7)	Mid-spectral (green or yellow) or grayscale are preferable to red or blue	—
Color Contrast (Secs 3.3 and 4.7)	High color contrast can be equivalent to high luminance contrast	High color contrast is not as good as high luminance contrast
Text Luminance (Sec. 4.7)	Try to achieve 100 cd/m^2 for young people, and up to 300 cd/m^2 for older people	Try to achieve 500 cd/m^2
Contrast Polarity of Text: light-on-dark or dark-on-light (Sec. 3.3)	Both polarities work	Some do better with light-on-dark.

5

The MNREAD Acuity Chart

J. Stephen Mansfield & Gordon E. Legge

5.1 INTRODUCTION

The MNREAD acuity chart is a clinical and research instrument for assessing how a person's reading performance is affected by print size.

Previous research has shown that standard clinical measures of letter acuity, contrast sensitivity, and visual field are poor predictors of overall reading performance (R12,1992; R13, 1995). Knowing a patient's Snellen acuity, field and ocular-media status, diagnosis, and age does not provide sufficient information for a satisfactory evaluation of reading potential. This should not be surprising because reading is a complex skill that relies on more than simply having good vision. Reading involves the combination of sensory abilities (such as visual acuity and contrast sensitivity), motor skills (to control reading eye movements) and cognitive abilities (such as vocabulary, language and reading skill). It is difficult to predict how these factors interact to govern reading performance. Thus the best way to assess the impact of visual factors on reading performance is to measure reading performance directly.

In R8 (1989) we introduced the Minnesota Low-Vision Reading Test, a computer-based system that uses a flashcard method for measuring reading speed (see section 2.1). This test was later simplified to use printed cards rather than a computer display (Ahn et al., 1995). This test (called the MNREAD) was designed to measure reading speed at large print sizes, where performance is not affected by the reader's acuity limit. At the same time we started to develop the MNREAD acuity chart. This chart used the same sentences as the MNREAD test, but the sentences were printed at a range of print sizes. We presented the MNREAD acuity charts at the OSA Noninvasive Assessment of the Visual System conference in Monterey in 1993 (Mansfield, Ahn, Legge, & Luebker, 1993). At that meeting we were persuaded to consider making charts that used a proportionally spaced font, such as Times Roman, that would be more representative of the print found in everyday reading material. This directly led to the development of the MNREAD acuity chart in its current form (see Fig. 5.1). Since then, versions of the MNREAD acuity chart have been developed in Japanese, Italian, French and Portuguese (see Fig. 5.2).

The MNREAD Acuity Chart

The MNREAD acuity chart is a continuous text reading acuity chart for normal and low vision. It is intended to have a wide range of applications. Clinical applications include prescribing optical corrections for reading or other near tasks, low vision assessment, and prescription of magnifiers or other reading aids. Research applications include evaluation of the effect of eye treatment or therapies on reading vision, and assessment of environmental viewing conditions (e.g., illumination) on reading. The chart may also be used for pediatric applications including the assessment of visual reading in children with low vision.

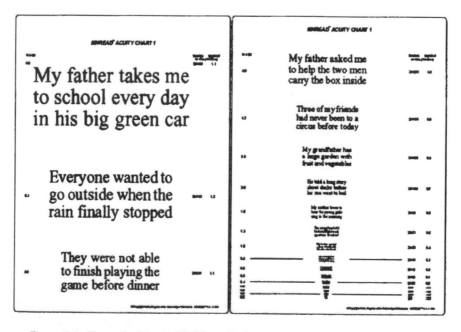

Figure 5.1   Example of the MNREAD acuity chart. The actual chart measures 11 × 14 in., and is printed with a resolution of 3,600 dpi.

Figure 5.2 Examples of foreign language versions of the MNREAD chart: (a) Italian, (b) French, (c) Portuguese, (d) Japanese, and (e) Japanese random word chart. We collaborated with the following colleagues to develop these charts—Italian: Gianni Virgili (University of Florence, Italy); French: Marie-Josée Senecal (Institut Nazareth et Louis-Braille, Canada), Jaques Gresset (Université de Montréal, Canada), Olga Overbury (McGill Low-Vision Center, Canada); Portuguese: Celina Tamaki (Universidade Federal de São Paulo, Brazil); and Japanese: Koichi Oda (Tokyo Women's Christian University, Japan). Development of these versions of the chart required much more than simple translation. These versions have their own sentences and meet the design principles described in section 5.2. To be consistent with everyday reading material, the text in the Japanese versions is presented in vertical columns: Each column is read top to bottom starting at the right side of the page.

Each MNREAD chart consists of a series of sentences printed in a sequence of decreasing print size. Testing results in a plot showing how a person's reading performance depends on print size (see Fig. 5.3).

Because all the sentences have the same number of characters and the same spatial layout, changes in reading performance for the different sentences are primarily due to the different print sizes. Plots of reading speed versus print size reveal three functional measures of reading performance:

1. Reading acuity: the smallest print that can just be read.

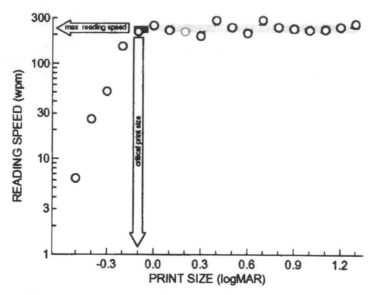

Figure 5.3 Reading speed versus print size data from a participant with normal vision. The horizontal gray bar represents the reading-speed plateau: the range of print sizes where the participant read with his maximum reading speed (220 wpm). The smallest print size on this plateau is the critical print size (CPS; –0.1 logMAR). Reading slows down for print sizes smaller than the CPS. The smallest print read is the participant's reading acuity. Section 5.4 and Box 5.1 describe the calculation of reading acuity, maximum reading speed and CPS.

2. Maximum reading speed: the reading speed when performance is not limited by print size.
3. Critical print size (CPS): the smallest print that supports the maximum reading speed.

These measures are important in the assessment of how vision affects reading ability.

Reading acuity is the tiniest print that can just be read. This measure indicates the absolute limit on reading small print. With print at the acuity size, a person will be able to identify most of the letters in the words (although reading may be slow and disjointed). Reading acuity closely corresponds to visual acuity as measured with a conventional letter acuity chart. Although the numerical values of reading acuity and letter acuity may differ, they are expected to be highly correlated. For example, in R15 (1996) the correlation between reading acuity measured with the MNREAD charts and letter acuity measured with the near Lighthouse Early Treatment of Diabetic Retinopathy Study (ETDRS) letter acuity chart was 0.94.[1] Compared to letter

[1] Also, in a sample of 56 individuals (21 with low vision and 35 with normal vision), Mansfield, Ahn, Legge, and Luebker (1993) found a correlation of 0.97 between reading acuity measured with the MNREAD charts and letter acuity measured with the near Lighthouse ETDRS letter acuity chart.

acuity, reading acuity is a more functionally relevant measure of vision in that it includes the effects of the cognitive and visual factors that are involved in a normal reading task: the effects of context (in which the reader may be able to determine letter and word identities based on the context of the other words that have already been read in the sentence), and the effects of crowding of nearby letters and words either beside or above and below the word being read (which may make the reading task more difficult).

Maximum reading speed is the best reading performance that can be attained when print size is not a limiting factor. This speed is a realistic goal for rehabilitation for patients with low vision. For many patients, maximum reading speed may only be possible with print sizes that are much larger than sizes normally encountered in everyday situations. One goal of rehabilitation is to use strategies enabling the patient to achieve maximum reading speed for everyday reading materials. Section 4.6 reviews strategies for achieving adequate magnification.

The CPS can be used to calculate the lowest power of magnification that will allow a patient to read at his or her maximum reading speed. It is an indicator of the optimum magnification required by a patient. With magnifiers there is often a tradeoff between how much the text is magnified and the size of the field of view. As text is magnified, fewer letters fit into the field of view. Restricting how many letters can be seen at a time impacts reading performance (R1, 1985; R2, 1985) and page navigation (R14, 1996). Thus the goal of magnification ought to be to magnify the text just enough so that the patient can read at his or her maximum reading rate. This size is the CPS. The CPS can also be used to determine the minimum print size (or maximum viewing distance) for children in schools, or for other settings where reading performance is functionally important but where there are constraints on either the size of a text display (e.g., on a mobile phone) or on the viewing distance (e.g., in the cockpit of a plane).

Note that laboratory measurements of reading speed versus print size generally show a reduction in reading speed at very large print sizes (R1, 1985; R2, 1985). However, the range of character sizes on the MNREAD acuity chart is usually insufficient to reach this large-print downturn. As a result, data from the chart are usually interpreted as showing a plateau in reading speeds for characters larger than the CPS. Nevertheless, the large-print downturn is sometimes reached when the chart is used at very short viewing distances, or when the patient reads with a severely restricted field of view. For example, Figure 5.4 shows MNREAD data from a patient with macular degeneration who has a scotoma in central vision that contains an island of preserved vision. For this patient there is only a small range of print sizes that supports relatively fast reading.

5.2 DESIGN PRINCIPLES

The overall goal was to design continuous text reading acuity charts using text passages that closely resemble normal everyday reading. Successful reading requires the dynamic integration of perceptual processes (letter and word recognition),

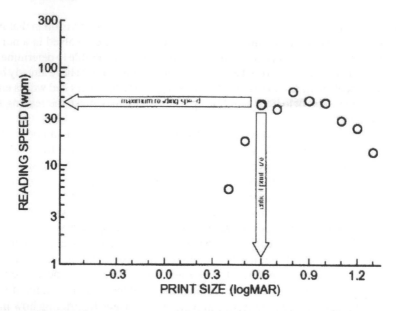

Figure 5.4   Reading speed versus print size data from a participant with a ring scotoma (central vision loss, but with an island of intact retina where vision is preserved). This plot shows the small range of print sizes that supports the patient's maximum reading speed.

oculomotor control (eye movements along lines of text and up and down the page), and higher cognition (use of semantic and syntactic cues). Each of these facets of reading should be involved in the reading test in order for the chart to be a valid predictor of everyday reading performance. Another design goal is that the visual characteristics of the chart conform with design principles for modern visual acuity charts (Bailey & Lovie, 1976; National Research Council, 1980). These design principles allow for comparable results for testing at different viewing distances, and allow for a precise measure of reading acuity.

Print Size and Resolution

The definition of print size is fundamental to any eye chart. The characteristic of the print used to define print size should be clearly specified. Print size on most letter acuity charts is unambiguous because the optotypes, such as the Sloan letters used on some modern charts, are designed to have the same height. However, continuous text involves a mixture of upper and lowercase letters, some with ascenders (b, d, f, h, k, l) and some with descenders (g, j, p, q, y). Thus, merely specifying letter height is insufficient. For Latin alphabets the height of a lowercase letter that has no ascenders or descenders should be used. For non-Latin alphabets the height of a common, well-known character should be used, one that has an unambiguous height. The MNREAD acuity chart defines print size according to the height of a lower case "x." For example, the font size of letters on the 20/20 line of the chart is scaled so that the letter "x" is precisely 0.0582 cm in height, so that it subtends 5 min-arc at a viewing distance of 40 cm.

The range of print sizes on an eye chart is also an important factor. The smallest print size should be smaller than the acuity limit of people with normal vision when the chart is viewed from the standard testing distance. This is important to ensure that an acuity threshold can be measured for all patients. Many young adults have acuities that are substantially better than 20/20 (0.0 logMAR), and some can achieve 20/8 (−0.4 logMAR). By including a size which is likely to be below threshold for all readers the chart avoids a ceiling effect. The largest print size should be as large as practical, to accommodate the widest range of low-vision participants.

MNREAD has print sizes from logMAR −0.5 to 1.3 (corresponding to Snellen 20/6.3 to 20/400) at the standard reading distance of 40 cm. Patients with low acuity can be tested at shorter viewing distances, thereby shifting the range of testable print sizes to a higher range. For instance, by using the chart with a 10 cm viewing distance the largest print size on the chart corresponds to 1.9 logMAR (20/1600).

The print size steps between successive sentences should follow a logarithmic progression. In this way, the size ratio between adjacent sentences is the same regardless of viewing distance. MNREAD uses 0.1 logMAR steps (i.e., the x-height in each sentence is 1.259 times larger than in the sentence below it) similar to many modern acuity charts such as the Bailey-Lovie visual acuity charts (Bailey & Lovie, 1976) and the Lighthouse Near and Distance EDTRS letter acuity charts.

Print size should be specified on the chart in logMAR, and in any other well-defined units that are relevant for the likely users. The English versions of the MNREAD use Snellen and M notation in addition to logMAR. See Appendix A for discussion of these units and for conversion formulas.

Typeface Properties

The goal of the chart is to mimic everyday reading material so it is important that key properties of the print (i.e., the typeface, character-to-character spacing, and line-to-line spacing) on the chart are similar to those used for everyday printed material such as newspapers, magazines, books, and so forth. Similarly, accents on letters should be used where they naturally occur in text (intentionally avoiding accented letters might unduly distort the representativeness of the text).

Typography varies in different cultures. For example, in the USA and Europe the typefaces used in books, magazines, and newspapers usually have serifs and are proportionally spaced (like Times Roman). In Japan, fixed-width typefaces are used for the Kanji and Hira-Kana scripts. In R15 (1996; see sections 4.1 & 4.2) we found that typeface had a significant impact on reading performance, especially for print sizes close to the acuity limit. Therefore, it is important that the typeface on the eye charts matches that which is used for the patients' everyday reading. Accordingly, the English, Italian, French, and Portuguese versions of the MNREAD chart use the Times Roman font, and the Japanese versions use the Heisei-Mincho typeface (which is similar to that used in Japanese newspapers and books).

One important deviation from the requirement to mimic everyday reading material concerns typographic scaling. It is conventional typographical practice for proportionally broader and bolder letters to be used for smaller print sizes. However, this should be avoided for a reading acuity chart so that the characters at small print sizes are scaled replicas of the characters used at larger sizes. Typographic scaling is undesirable because if smaller characters were made proportionally broader then the typographic properties of the chart would vary depending on viewing distance (letters with an x-height of 5 min-arc viewed from 20 cm would be broader than 5 min-arc letters viewed from 40 cm). Moreover, even for a fixed viewing distance, it would be unclear whether any change in reading performance found with small print is due to the smaller size or due to the difference in stroke width.

A final typographic concern is that the chart is printed with sufficient resolution so that coarse sampling will not compromise acuity assessment. The MNREAD charts have been produced using a photo typesetting method with a resolution of 3600 dots/inch. For the smallest print on the chart (–0.5 logMAR), this corresponds to about 20 dots per x-height, a sampling density that is adequate for good legibility (see R1, 1985, and section 4.3).

Text Properties

To equate the reading requirements at each print size, all the sentences on the chart must be matched for length. Sentence length is best measured as the number of characters (this count includes the spaces between words and an implied period at the end of the sentence.) The selection of sentence length (i.e., the number of characters per sentence) represents a compromise between: (a) having sufficient text to estimate reading performance, (b) having too much text to fit on the chart at the large-print end, and (c) the time required by low-vision participants to read the sentences. MNREAD uses 60 characters per sentence. This turns out to be convenient for scoring if we define a "standard-length word" to have 6 characters (see Carver, 1990). A 60-character sentence consists of 10 standard-length words. Using standard-length words helps minimize the differences in scoring that would occur due to the different word lengths found in different sentences. For example, some of the sentences have 13 relatively short words, whereas others have just 10 words of a longer length.

It is also important that all the sentences on the chart be matched for spatial layout. All sentences should fit snugly into a rectangular box of fixed aspect ratio. The size scaling of the sentences on successive lines is equivalent to scaling the size of this bounding box. The sentences should be printed onto three or more lines of text so that there will be at least one line with text above and below it to include the potential impact of vertical crowding on reading. The lines of text should be left and right justified so that each sentence fits into the bounding rectangle. To allow for complete justification when proportionally spaced fonts are used, the white space between words can be adjusted. However, to ensure uniformity of word-to-word spacing the total amount of padding should not exceed 0.5 mean character widths. If a fixed-width font is used no padding is needed.

MNREAD uses 60 characters on 3 lines. The bounding box for each sentence is 17.53 x-heights wide (based on the average width of 19 Times Roman letters), and 5.49 x-heights tall (ignoring ascenders and descenders—this is the height of 3 lines of text using normal line-to-line spacing in the Times Roman font). Because Times Roman is a proportional font, the spatial constraints are sometimes met by sentences that have unequal numbers of characters on the 3 lines (e.g., 21, 19, 20). A challenge in constructing MNREAD sentences is to simultaneously meet the 60-character constraint and the bounding-box spatial constraint.

Sentence Composition

The sentences should be simple declarative sentences. The sentences should be independent from each other in their semantic content. There should be no continuity in story line or theme between the sentences. In this way, reading the linguistic content of an earlier sentence will have no impact on the reading performance of subsequent sentences. If the sentences combined to create a story, performance at smaller print sizes may be enhanced due to the ability to use the thematic context from previous sentences to help decode hard-to-see words.

The vocabulary should be selected from high-frequency words in reading material for 8-year-old children (3rd-grade students in the United States). Restricting the chart to a relatively simple vocabulary extends the range of reading abilities that can be assessed using the chart, reducing the chances that the words will be unfamiliar to the patient. Carver (1990) showed that reading performance is unaffected by text difficulty provided that the reading material is substantially simpler than the reader's grade level. The English version of the MNREAD was based on a restricted vocabulary of this type. The majority of the words are among the 1000 most frequent words in third grade school books.

In selecting words for the MNREAD charts we avoided using hyphenated words and contractions. Such words complicate scoring procedures due to the ambiguity of whether they count as one or two words. We also avoided words with regional spellings (such as American "color" vs. British "colour"), or with regional meanings (e.g., American "pants" vs. British "trousers"). This is especially important for languages that are spoken in several countries.

A further constraint in the choice of words is to minimize the frequent repetition of concrete words (word selection is often restricted by the combined constraints of using a limited vocabulary and of fitting the sentence into the prescribed length and format). For example in an early version of the Portuguese chart the term for "child" occurred in many of the initial set of 19 sentences. This is undesirable because it may make it easier for readers to guess correctly when they are unsure of a word.

In selecting sentences for the MNREAD charts, sentences are chosen to be similar to each other in their typical reading time (based on pilot testing with normally sighted participants). This eliminates awkward tongue-twister sentences that are difficult to say, and which could artificially inflate reading times.

A potential problem that comes with using sentences on a reading test is that patients may remember the sentences from earlier testing administrations. There should be at least two versions of the chart, for left and right eye testing, or pre- and posttreatment testing. Additional versions may be useful for applications in which the same participant is tested repeatedly. Each MNREAD chart is produced with at least two versions using different sentences. We have an additional pool of sentences for research purposes.

Physical Properties of the Charts

To ensure that reading performance on the eye chart is limited primarily by print size, the text on the chart should have high luminance contrast (> 80% Michelson definition). R5 (1987) and R6 (1989) showed that reading speed in participants with normal vision can tolerate substantial contrast reduction, but that reading speed in some patients with low vision is severely reduced by even modest contrast reduction. The MNREAD charts have contrast exceeding 85%. We have also produced sets of charts for testing the contrast dependence of reading (O'Brien et al., 2000).

Some low-vision patients (especially those with cloudy ocular media) have better reading performance with reversed contrast polarity (white text on a black background) than with normal contrast polarity (see section 3.3). The MNREAD chart is available both as a black-on-white chart, and as a white-on-black chart for testing this possibility.

Finally, in order to mimic everyday reading it is important that the physical material that the chart is printed on has a matte surface (like newspaper). A matte surface also eliminates the possibility that specular reflections will interfere with the reading task.

5.3 GUIDELINES FOR USING THE MNREAD ACUITY CHARTS

Conditions for Testing

The chart should be lit evenly so that no shadows or glare will interfere with reading. The luminance of the white background on the charts should be 100 cd/m². This is usually achievable using ordinary room lighting or desk lamps.

The print sizes and markings on the chart are designed for testing at the standard reading distance of 40 cm (16 in.). However, the range of print sizes can be extended to larger values for low vision by testing at a shorter viewing distance. A head rest can be used to maintain a constant viewing distance to the chart. This will prevent the patient from creeping forward to see the smaller print. If a head rest is used with patients who have central field loss, they may need to position the MNREAD card themselves so that the sentences can be comfortably read with their peripheral vision.

Testing Procedure

Patients should be instructed as follows:

> When I say "start" please read the sentence aloud as quickly as you can with out making errors. But if you do make an error, or realize that you have missed a word, continue to read to the end of the sentence and then go back and correct yourself.

Start with the largest sentence and move onto the subsequent sentences in decreasing size order. Keep going until the patient cannot read any words in a sentence. Note any reading errors (i.e., any words that are missed or read incorrectly such as reading "dog" instead of "dogs"), and note the time taken to read the sentence to the nearest 0.1 sec. Reading time is defined as the time between when the patient is told to "start" and when the patient finishes uttering the last word in the sentence. If a patient makes an error on a word, and then goes back and corrects that word *before* reaching the end of the sentence that error should not be counted.

During testing it is useful for the examiner to use a blank card to cover the sentence that is about to be read. The sentence should be uncovered at the same time as the examiner instructs the patient to start reading. This gives an objective means to determine when the reading process starts, and prevents the patient from previewing the next sentence.

5.4 SCORING GUIDELINES

Calculating Reading Acuity

A quick estimate of reading acuity is given by the smallest print size at which the patient can read the entire sentence without making significant errors. (Usually reading performance deteriorates rapidly as the acuity limit is approached, and it is easy to determine the level where reading becomes impossible). This method measures acuity to the nearest 0.1 logMAR. If the chart was used at a nonstandard distance (i.e., anything other than 40 cm), the acuity estimate needs to be adjusted to account for the distance used (see Table 5.1).

A more sensitive measure of reading acuity can be determined by counting the number of words read incorrectly. This scoring procedure is equivalent to the "letter-by-letter" method described by Ferris, Kassoff, Bresnick, and Bailey (1982) for scoring visual acuity with letter charts. The number of reading errors can be used to

Table 5.1
Print Size Adjustment to logMAR Print Size for Different Viewing Distances

Viewing distance		Correction
cm	inches	logMAR
4	1.6	+1.00
8	3.1	+0.70
12	4.7	+0.52
16	6.3	+0.40
20	7.9	+0.30
24	9.4	+0.22
28	11.0	+0.15
32	12.6	+0.10
36	14.2	+0.05
40	15.7	+0.00
44	17.3	−0.04
48	18.9	−0.52
52	20.5	−0.56
56	22.0	−0.60
60	23.6	−0.18
64	25.2	−0.20
68	26.8	−0.23
72	28.3	−0.26
76	29.9	−0.28
80	31.5	−0.30

interpolate within the 0.1 logMAR steps on the chart. Each sentence consists of 10 standard-length words (see section 5.2) which effectively divide the 0.1 logMAR increment in 10 smaller steps of 0.01 logMAR. After the patient has read as much of the chart as possible, note the print size (in logMAR) of the last *sentence* attempted, and count the number of *words* that the patient read incorrectly from all sentences on the chart (If there are more than 10 errors in a single sentence only count that as 10.) Calculate reading acuity (in logMAR) using the following formula:

$$\text{Acuity} = \text{smallest print size attempted} + (\text{word errors} \times 0.01). \qquad (5.1)$$

If the chart was used at a nonstandard distance (i.e., anything other than 40 cm), the acuity estimate needs to be adjusted to account for the distance used (see Table 5.1). An example of this calculation is shown in Box 5.1.

Box 5.1

Sample Calculation of Reading Acuity and Reading Speed

Figure 5.5 shows example data obtained with the MNREAD acuity chart. The patient was tested using a viewing distance of 32 cm. Starting from the 1.3 logMAR level, the patient read all or part of 19 sentences. On the score sheet the tester has drawn a line through each word that was missed or read incorrectly, and has noted the reading time (in seconds) beside each sentence. From the score sheet it can be seen that the patient made a total of 10 reading errors.

The patient's reading acuity is calculated using Equation 5.1 as follows:

$$\begin{aligned}
\text{Acuity} &= \text{size of last sentence attempted} + (\text{errors} \times 0.01) \\
&= -0.5 + 10 \times 0.01 \\
&= -0.5 + 0.1 \\
&= -0.4 \text{ logMAR.}
\end{aligned}$$

This value now needs to be corrected for the non-standard viewing distance that was used. Table 5.1 shows that for a 32 cm viewing distance the print size should be adjusted by +0.1 logMAR. Thus, the patient's reading acuity is $-0.4 + 0.1 = -.3$ logMAR.

The patient's reading speed for each sentence can be calculated using Equations 5.2 or 5.3. For example, the patient read the 1.1 logMAR sentence in 2.4 sec, but made one reading error.

Ignoring the error, reading speed is calculated using Equation 5.2:

$$\begin{aligned}
\text{Reading speed (wpm)} &= 600/(\text{reading time in seconds}) \\
&= 600/2.4 \\
&= 250 \text{ wpm}
\end{aligned}$$

Taking the reading error into account, reading speed is calculated using Equation 5.3:

$$\begin{aligned}
\text{Reading speed (wpm)} &= 60 \times (10 - \text{errors})/(\text{reading time in seconds}) \\
&= 60 \times 9/2.4 \\
&= 225 \text{ wpm}
\end{aligned}$$

The MNREAD data sheet (in Fig. 5.5, also available as a clean copy on the accompanying compact disk) is designed to simplify the reading-speed calculations required to create a graph of the patient's reading speed as a function of print size. The axis at the top of the data plot shows reading time (in seconds). The axis at the bottom of the plot uses Equation 5.2 to convert the reading times into reading speeds (in wpm). To create the reading speed by print size graph it is simple to plot the reading times for each sentence using the top reading-time axis.

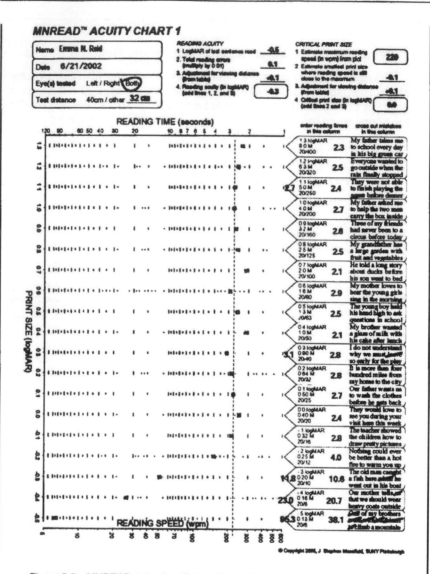

Figure 5.5 MNREAD data sheet for a patient with normal vision, showing reading errors, reading times, reading-speed versus print-size graph, and the calculation of reading acuity, maximum reading speed and critical print size. A clean copy of this data sheet is available on the accompanying compact disk.

The data plot can also be used to plot reading speeds taking reading errors into account. However, rather than directly using Equation 5.3 to calculate reading speed, it is easier to adjust the reading times for any errors that were made. For each sentence count the errors and multiply the reading time by the factor specified in Table 5.2. For example the patient read the 1.1 logMAR sentence in 2.4 sec but made one error: the adjusted reading time should be 2.4 × 1.11 = 2.7 sec (i.e., the patient read 9 words in 2.4 sec, which is the same speed as reading 10 words in 2.7 sec). The data sheet in Figure 5.5 shows the adjusted reading times to the left of each sentence that contained a reading error. These adjusted times are plotted.

Table 5.2

Reading Time Correction for Sentences with Reading Errors (for use when Plotting Reading Time Data on the MNREAD Data Sheet)

Number of errors	Correction
0	× 1
1	× 1.11
2	× 1.25
3	× 1.43
4	× 1.67
5	× 2
6	× 2.5
7	× 3.33
8	× 5
9	× 10
more than 9	× 0

The plot shows that, for print sizes larger than 0.0 or –0.1 logMAR, the patient's reading speed was approximately constant at about 220 wpm. The tester has drawn in a dashed line at this speed, this indicates the patient's maximum reading speed.

When the print size was smaller than –0.1 logMAR the patient's reading speed deteriorated. It can be seen that –0.1 logMAR was the smallest print size that could be read close to the maximum reading rate: –0.1 logMAR is the critical print size. This value needs to be corrected for the nonstandard viewing distance that was used. Table 5.1 shows that when the chart is viewed from 32 cm the reading acuity value needs to be adjusted by +0.1 logMAR. Thus, the patient's critical print size is –0.1 + 0.1 = 0.0 logMAR.

Box 5.2 shows computer programs that can be used to automate the calculation of critical print size and maximum reading speed, as an alternative to estimating these parameters by eye. (The computer source code is also available on the accompanying compact disk.) Nevertheless, we recommend that the results from the computer algorithm be visually checked against a graphical plot of the MNREAD data.

Calculation of Reading Speed

Reading speed is measured in words per minute (wpm). With the MNREAD acuity charts the reading speed calculation is simplified because each sentence has the same length: 10 standard length words. Reading speed is given by:

$$\text{reading speed (wpm)} = 600 \,/\, (\text{time in seconds}) \tag{5.2}$$

A more precise reading speed measurement can be achieved by excluding words that were missed or read incorrectly. In this case reading speed is given by:

$$\text{reading speed (wpm)} = 60 \times (10 - \text{errors}) \,/\, (\text{time in seconds}) \tag{5.3}$$

If 10 or more errors were made in a sentence then the reading speed for that sentence can be assumed to be zero. An example of this calculation is shown in Box 5.1.

CPS and Maximum Reading Speed

The simplest means to determine CPS and maximum reading speed is to visually examine the MNREAD data on a plot of log reading speed[2] as a function of print size. Typically such plots show approximately constant reading speed over a range of large print sizes (a reading-speed plateau) with a decline in reading speed at smaller print sizes.

We define the maximum reading speed to be the average reading speed on the reading-speed plateau. We define the CPS to be the smallest print size on this reading-speed plateau. The maximum reading speed plateau and CPS are usually easy to determine by eye, especially for participants with normal vision, who typically show only a small variation in reading speed at large sizes.

Some studies have estimated the CPS using curve-fitting procedures to characterize the reading-speed versus print-size function with a bilinear fit, a compressive exponential curve, or other convenient functions (e.g., Latham & Whitaker, 1996; R18, 1998). The curve fitting approach works well if there are multiple reading-speed estimates at each print size so that the variability associated with each measure can be used to weight the data points appropriately. However, this is generally not the case for typical MNREAD data in which each data point is usually based on the patient reading only one sentence. In R15 (1996) we argued that the variability in reading speed at small print sizes can significantly affect the estimation of the CPS using a bilinear fit. To overcome this problem we developed a computer algorithm to automate the calculation of the CPS and maximum reading speed (source code for this algorithm in Perl, MATLAB, and C is given in Box 5.2 and also on the accompanying compact disk). The algorithm searches for a read-

---

[2]Typically, variability in reading speed measurements is proportional to the overall reading speed. Therefore, reading speed should be plotted using a logarithmic scale (so that the variability is constant at all speeds).

ing-speed plateau in the MNREAD data. A plateau is defined as a range of print sizes that supports reading speed at a significantly faster rate than the print sizes smaller or larger than the plateau range. This is determined by calculating the average and standard deviation of the reading speeds on the plateau, and checking that the reading speed at the other print sizes is at least 1.96 SDs slower than the average speed on the plateau. Generally several print-size ranges will qualify as reading-speed plateaus based on this criterion. The algorithm chooses the plateau with the fastest average reading speed. The smallest print size on the plateau is then taken as the CPS, and the average reading speed on the plateau is taken as the maximum reading speed.

For most MNREAD data this algorithm determines a CPS that is in close agreement with the estimates obtained by naïve raters who determine the CPS by eye. Nevertheless, there are occasions where the algorithm will provide an estimate that is clearly wrong. For example, if there is very little variability in the reading speeds measured for large print sizes the algorithm locates the CPS at the slightest drop in reading speed. Thus we recommend that the results from the computer algorithm be visually checked against a graphical plot of the MNREAD data.

Kallie, Cheung, Legge, Owsley, and McGwin (2005) have recently investigated the use of nonlinear mixed effects (NLME) modeling to assist in curve fitting with MNREAD data. NLME is particularly promising because it uses parameter estimates from a larger group to allow suitable curve fits for individual data sets that contain few data points. Kallie et al. showed that using NLME modeling can significantly improve the performance of curve-fitting (using a bilinear fit, or a compressive exponential curve) for estimating CPS and maximum reading speed.

Reliability and Validity

The reliability of the MNREAD chart measures has been calculated as a part of our study of font effects on reading (R15, 1996), and more recently using the Italian version (Virgili et al., 2004). In R15 (1996) we tested 50 college students with normal vision on the two forms of the standard Times Roman MNREAD acuity chart. These data allowed us to assess the parallel-forms reliability for reading acuity, CPS, and maximum reading speed. The differences between each participant's scores on the two charts makes a distribution with a mean that is very close to zero and a standard deviation that represents how similar the scores tend to be on the two charts. Virgili et al. used a similar procedure with 116 school children with normal vision. The 95% confidence intervals (1.96 SDs of the difference distribution) are shown in Table 5.3.

Note that the confidence intervals for maximum reading speed correspond to an 18.3% change in reading speed for the R15 (1996) data, and a 19.4% change in reading speed for the Virgili et al. (2004) data. These values are approximately half the size of the confidence interval for measuring reading speed using a single large-print MNREAD sentence (Ahn et al., 1995).

The validity of the MNREAD measure of reading speed has been demonstrated by comparing reading speed with MNREAD sentences to everyday reading tasks. In R13 (1995), the reading speeds from 40 low-vision participants who read passages of printed text on paper with their preferred magnifiers were highly correlated ($r = 0.89$) with their reading speeds assessed using the flashcard method (without magnifier). This finding demonstrated that reading speed measured with large print MNREAD sentences is predictive of real-world reading with a magnifier. Recently, Tamaki et al. (2004) reported that maximum reading speeds measured from 20 participants with normal vision with the Portuguese version of the MNREAD chart were highly correlated ($r = 0.82$) with reading speeds for a short newspaper article.

5.5 CONCLUDING REMARKS

As with all clinical tests there are some limitations to the use of the MNREAD charts.

The measure of reading performance is based on reading short sentences. Using 10-word sentences is convenient for clinical testing because the reading data can be obtained relatively quickly. However, in many real reading situations (such as reading a novel, or the Sunday newspaper, or working at a computer) the reading task occurs over a long period of time. The maximum reading speed and CPS measured with the MNREAD chart may not be sustainable for long periods of reading. For example, the maximum reading speed with short sentences may overestimate the reading speed achieved with normal prolonged reading (in the same way that sprinters run faster than marathon runners).

Another limitation is that reading speed measured with the MNREAD test is constrained by vocalization rate. Having patients read aloud is convenient for clinical testing, as it allows objective scoring. For patients reading close to their acuity limit, reading performance is likely to be slow, whether the text is read aloud or silent. However, for participants with normal vision, it is likely that text can be read faster than it can be spoken. In these circumstances, the maximum reading speed measured with the chart may be slower than the participant's actual maximum reading speed.

Despite these limitations, the reliability and validity data presented above and in Table 5.3 indicate that the MNREAD chart provides a straightforward and reliable assessment of real world reading function. Overall, the MNREAD acuity chart is a

Table 5.3

Confidence Intervals for MNREAD Acuity Chart
(Based on data from R15, 1986; and Virgili et al., 2004)

Measure	95% Confidence Intervals	
	R15 (1986)	Virgili et al. (2004)
Reading Acuity:	±0.12 logMAR	±0.14 logMAR
Critical Print Size:	±0.25 logMAR	±0.19 logMAR
Maximum Reading Speed:	±0.073 logWPM	±0.077 logWPM

practical application of a wide range of the research described in the Psychophysics of Reading series.

Box 5.2
Source code

(This source code is also provided on the accompanying compact disk.)
Source code in Perl

```perl
#! /usr/bin/perl
# mnreadfit
use warnings;
our @data;

#read data from stdin
while(<>){
 my($size,$time,$errors)=split;
 $errors||=0;
 push @data, {SIZE=>$size, LWPM=>log(60*(10-$errors)/$time)}
   unless $time<=0 or $errors>9;
}


my $fit = mnfit(\@data); #find the reading speed plateau

#print the results
printf "CPS: %.1f logMAR\nMRS: %.1f wpm\n",$fit->{CPS},$fit->{MRS};
sub mnfit {          #routine to calculate the CPS and MRS
 my $aref=shift;
 my ($n,$top) = (@$aref-1,{LWPM=>-1});

 @$aref = sort {$a->{SIZE}<=>$b->{SIZE}} @$aref;

 for my $sp (0..$n-1){
  for my $bp ($sp+1..$n){
    my($amax,$pmean,$psd) = pstats($aref,$sp,$bp);
    $top = {SMALL=>$sp,BIG=>$bp,LWPM=>$pmean}
     if $amax<$pmean-1.95*$psd and $pmean>$top->{LWPM};
  }
 }
```

```
  return {MRS=>exp($top->{LWPM}),CPS=>$aref->[$top->{SMALL}]{SIZE}};
}


sub pstats {
 my($aref,$sp,$bp)=@_;
 my ($n,$max,$s,$ss,$sn)=(@$aref-1,-1000);

 for my $i (0..$n){
  my $pt = $$aref[$i];
  $max=$pt->{LWPM}>$max ? $pt->{LWPM} : $max and next if
   $i<$sp or $i>$bp;
  $s +=$pt->{LWPM};
  $ss+=$pt->{LWPM}*$pt->{LWPM};
  $sn++;
 }
 return ($max,$s/$sn,sqrt(($ss-$s*$s/$sn)/($sn-1)));
}
```

This file should be named: mnreadfit.

Example

The above program reads the data file specified on the command
line. Each line of the data file specifies (a) the logMAR print size, (b)
the reading time (in seconds) for that print size, and (c) the number of
reading errors for that print size (this third parameter is optional, and
defaults to zero). These parameters should be separated by spaces.
For example, save these data to a file named: data

```
1.3 5.0 1
1.2 4.6
1.1 4.5
1.0 4.2
0.9 4.1
0.8 4.3
0.6 4.5
0.5 4.0
0.4 4.1
0.3 4.2
0.1 4.1
```

```
0.0 4.3
-.1 4.3 1
-.2 4.7
-.3 10.5 1
-.4 30.1 3
```

Here is the output from the mnfit program:

```
mnreadfit < data
CPS: 0.0 logMAR
MRS: 142.9 wpm
```

For these data, the critical print size is 0.0 logMAR, and the maximum reading speed is 142.9 wpm.

Source code in MATLAB

```
function [rac,cps,lps,mrs,sd,rs] = mnreadfit(data);


smallest = 1.4;
fastest_mean = 0;
data(:,find(data(:,3)>10)) = 10;
total_errors = sum(data(:,3));


if min(data(:,1))<smallest;
smallest = min(data(:,1));
end;


data(:,4) = 60*(10-data(:,3))./data(:,2);
data(:,5) = log(data(:,4));


data = sortrows(data);


rs = data(:,4);


for i = 2:length(data)-2;
  for j = i+2:length(data);
  omax = max( data( find(data(:,1)<data(i,1) |
    data(:,1)>data(j,1) ),4) );
```

```
  p_mean = mean(data([i:j],5));
  p_std = std(data([i:j],5));
  if log(omax)<p_mean-1.96*p_std;
     if p_mean>fastest_mean;
        fastest_mean = p_mean;
        fastest_std = p_std;
        fastest_cps = data(i,1);
        fastest_lps = data(j,1);
     end;1
   end;
  end;
end;
```

Source code in C

```c
/*MNREADFIT.C*/
#include <stdio.h>
#include <math.h>

typedef struct _sentence_data {
  double logmar;
  double secs;
  int errors;
  double wpm;
  double lwpm;
} SENTENCE_DATA;

typedef struct _plateau {
  double mean;
  double stdev;
  double cps,lps;
} PLATEAU;

int is_a_plateau();
int compare_logmar();

main()
{
  SENTENCE_DATA data[19],*s;
```

```c
PLATEAU p,fastest;
double logmar, secs, smallest;
int errors,total_errors=0;
int a,b;
char line[256];
int nargs, npts=0;

/*read data from stdin*/
while(fgets(line,256,stdin)!=NULL){
  s=&data[npts];
  nargs=sscanf(line,"%lf %lf %d",&logmar,&secs,&errors);
  if(nargs<3) errors=0;
  if(secs>0 && errors<10){
    s=&data[npts];
    s->logmar = logmar;
    s->wpm = 60*(10-errors)/secs;
    s->lwpm = log(s->wpm);
  }
  npts++;

  total_errors+=((errors>10)?10:errors);
  }

/*sort the data by print size*/
qsort(&data,npts,sizeof(SENTENCE_DATA),&compare_logmar);

smallest=data[0].logmar; /*needed for RAC caclulation*/

/*find the maximum reading speed plateau*/
for(a=0;a<npts-2;a++) for(b=a+2;b<npts;b++){ /* set a & b to all
                                pairs of points */
 if(is_a_plateau(data,a,b,npts,&p)){ /* test if a->b is a plateau */
   if(p.mean>fastest.mean){ /* store range giving fastest mean */
     fastest.mean=p.mean;
     fastest.stdev=p.stdev;
     fastest.cps=data[a].logmar; /* cps is smallest size on plateau */
     fastest.lps=data[b].logmar; /* lps is largest size on plateau */
   }
 }
```

```
    }

  printf("Reading acuity = %+.21f logMAR\n",smallest+total_errors/100.);
  printf("Critical print size = %+.11f logMAR\nLPS = %+.11f logMAR\n",
     fastest.cps,fastest.lps);
  printf("Maximum reading speed = %-5.11f wpm\n sd = %.11f%%\n",
    exp(fastest.mean), 100.*exp(fastest.stdev)-100.);
}


int compare_logmar(a,b) /*used by qsort*/
SENTENCE_DATA *a,*b;
{
  if (a->logmar < b->logmar) return -1;
  if (a->logmar > b->logmar) return 1;
  return 0;
}


int is_a_plateau(d,a,b,n,p) /* do sizes (a->b) satisfy plateau criteria?*/
SENTENCE_DATA *d;
PLATEAU *p; /*pointer to struct to receive plateau statistics*/
int a,b,n; /*a=low end of range, b=high end, n=numer of points*/
{
  int i;
  double sum,ssq,num;
  double omax;

  sum=ssq=num=omax=0.0;

  for(i=0;i<n;i++){
   if(i<a||i>b){ /*calculate maximum speed outside region*/
      if(d[i].wpm>omax) omax=d[i].wpm;
   } else { /*calculate mean and stdev of the region*/
      sum+=d[i].lwpm;
      ssq+=d[i].lwpm*d[i].lwpm;
      num++;
   }
  }
```

```
p->mean = sum/num;
p->stdev = sqrt((ssq-(sum*sum)/num)/(num-1.0));


/*indicate if it's a plateau*/
return (log(omax)< p->mean - 1.96*p->stdev);
}
```

Appendix: Print Size Definitions and Conversions

J. Stephen Mansfield & Gordon E. Legge

Many measures of print size (character size) have been used in vision research and clinical vision, several in our series of articles on the psychophysics of reading. Table A1 presents conversion formulas and definitions for seven measures.

DEFINITIONS OF MEASURES IN TABLE A1

Physical Size

x-height: height of lowercase "x" for continuous text charts, or height of optotype for charts using entirely uppercase letters or symbols, measured with a ruler or equivalent device in units such as millimeters, centimeters, or inches.

Sloan M: refers to the physical height of the optotype that subtends 5 min-arc at the distance in meters indicated by the M value. For instance, 2M characters subtend 5 min-arc at a viewing distance of 2 m. Although viewing distance is invoked in this definition, Sloan M is a measure of physical print size.

Angular Size

Degrees: The visual angle in degrees of an object of size x, viewed from a distance d, is closely approximated by $57.3 \, x/d$.

MAR: Minimum angle of resolution, based upon requirement to resolve critical details that are 1/5th of the optotype size. The critical details are distinguishing features of the optotype such as the gap in Landolt C stimuli or stroke width of the Sloan letters.

logMAR: The base 10 logarithm of the minimum angle of resolution.

Snellen fraction: Numerator signifies viewing distance (i.e., 20 ft or 6 m), denominator is the viewing distance (in the same units) at which the optotype subtends 5 min-arc .

Decimal acuity: Expresses the Snellen fraction as a decimal. For instance, the decimal acuity associated with the Snellen fraction 20/80 is 0.25. Note that decimal acuity is rarely used to specify print size per se, but as a means for expressing the size of acuity letters in a test of visual acuity.

OTHER RELATED MEASURES

We briefly mention some other frequently encountered measures of print size:

Snellen equivalent: refers to angular character size measured at one viewing distance (e.g., 40 cm) equal in angular size to a specified Snellen fraction. For instance, the "Snellen equivalent" of a 20/200 letter at a viewing distance of 40 cm subtends 50 min-arc, the same angular subtense as a 20/200 letter at 20 ft. Of course, the x-height of the 20/200 letter at 20 ft will be much larger than the x-height of the 20/200 letter at 40 cm.

Point size: This is a physical measure of print size derived from usage in typography. There are three definitions of point size: (a) According to the traditional Anglo-American type founder's usage, dating from 1886, there are 72.27 points per inch. (b) Adobe has redefined the point in the Postscript language so that there are exactly 72 points per inch. (c) In the European Didot point system, dating from the 18th century, there are 67.58 points to the inch. Computer operating systems, such as Microsoft Windows and Apple Macintosh, usually adopt the Postscript definition. It is difficult, however, to anticipate how the software will render characters of a given nominal point size on any particular computer display. Moreover, the conversion from x-height to a font's point size will depend on the font, and may change with the font's point size. The most straightforward way of calibrating point size with x-height is to do a physical measurement of the x-height.

Center-to-center size: In a fixed-width text font, each character is allocated equal horizontal distance on the page. This fixed distance (equivalent to the horizontal distance between the centers of a pair of identical adjacent letters) will typically be larger than the x-height. There is no constant conversion factor between center-to-center spacing and x-height, so a calibration measurement is required. In some of the papers in our series, we defined angular character size in terms of center-to-center spacing, having taken viewing distance into account.

Jaeger numbers: These values designate print sizes on some near charts of reading acuity, but they are poorly standardized.

USING TABLE A1

The table requires that viewing distances and x-height are measured in centimeters. Recall that there are 2.54 cm in 1 in. Within the table, conversion between the

TABLE A1

Convert From	To: x-height (cm)	Sloan M	Degrees	Decimal acuity	MAR (min-arc)	logMAR	Snellen denominator
Physical measures							
x-height in cm (x)	–	$x/0.1454$	$57.3\,x/d$	$d/(687.5x)$	$687.5\,x/d$	$\log_{10}(687.5x/d)$	$687.5\,S_N\,x/d$
Sloan M (M)	$0.1454\,M$	–	$8.333\,M/d$	$d/(100\,M)$	$100\,M/d$	$\log_{10}(100M/d)$	$100\,S_N\,M/d$
Angular measures							
Degrees (α)	$\alpha d/57.3$	$\alpha d/8.333$	–	$1/(12\,\alpha)$	12α	$\log_{10}(12\alpha)$	$S_N\,12\,\alpha$
Decimal acuity (A)	$d/(687.5\,A)$	$d/(100\,A)$	$1/(12\,A)$	–	$1/A$	$-\log_{10}A$	$S_N\,/A$
MAR min-arc (μ)	$\mu d/687.5$	$\mu d/100$	$\mu/12$	$1/\mu$	–	$\log_{10}\mu$	$S_N\,\mu$
logMAR (L)	$10^{L}\,d/687.5$	$10^{L}\,d/100$	$10^{L}/12$	$1/10^{L}$	10_{L}	–	$S_N\,10_{L}$
Snellen fraction (S_N/S_D)	$dS_D/(687.5\,S_N)$	$dS_D/(100\,S_N)$	$S_D/(12\,S_N)$	S_N/S_D	S_D/S_N	$\log_{10}(S_D/S_N)$	–

physical measures (x-height and Sloan M) does not involve viewing distance, nor does conversion between angular measures. Viewing distance is only required in converting between physical size (x-height or Sloan M) and one of the measures of angular size.

The table is easy to use. Suppose, for example, you want to convert character size in degrees to logMAR. Suppose the angular size is 0.21°. Find the row in Table A1 for Degrees (α). Under the column for logMAR, the conversion is $\log_{10}(12\alpha) = \log_{10}(2.52) = 0.40$. As another example, suppose you want to convert an x-height of 0.145 cm (= 1.45 mm) at a viewing distance of 40 cm to angular size in degrees. From the first row of the table, the conversion is 57.3 x/d = 57.3 (0.145/40) = 0.208°.

These and other numerical examples are shown in Table A2.

Table A2

Physical Measures					
x-height (cm)	0.0582			0.1454	
Sloan M	0.40			0.11	
Angular Measures	@ 100 cm	@ 40 cm	@ 20cm	@ 100cm	@ 40cm	@ 20cm
Degrees	0.0333	0.08333	0.1667	0.08333	0.2083	0.4166
Decimal acuity	2.50	1.00	0.5	1.00	0.40	0.20
MAR (min-arc)	0.40	1.00	2.00	1.00	2.50	5.00
logMAR	-0.40	0.00	0.30	0.00	0.40	0.70
Snellen fraction	20/8	20/20	20/40	20/20	20/50	20/100
Snellen fraction	6/2.4	6/6	6/12	6/6	6/15	6/30

Glossary
of Technical Acronyms

AMD. Age-related macular degeneration.

CCTV Magnifier. Closed-circuit television magnifier.

cpd. Cycles per degree.

CSD. Critical sample density.

CSF. Contrast sensitivity function.

CPS. Critical print size.

ETDRS Acuity Chart.. Early Treatment of Diabetic Retinopathy Study acuity chart.

fMRI.. Functional magnetic resonance imaging.

JMD. Juvenile macular degeneration.

M Pathway. Magnocellular pathway.

MAR. Minimum angle of resolution.

MNREAD Acuity Chart. Reading (READ) acuity chart, developed at the University of Minnesota (MN).

P Pathway. Parvocellular pathway.

PRL. Preferred retinal locus.

RP.. Retinitis pigmentosa.

RSVP.. Rapid serial visual presentation.

SLO. Scanning laser ophthalmoscope.

slwpm. Standard-length words per minute.

wpm. Words per minute.

References

Abdelmour, O., & Kalloniatis, M. (2001). Word acuity threshold as a function of contrast and retinal eccentricity. *Optometry and Vision Science, 78,* 914–919.

Abell, A. M. (1894). Rapid reading: Advantages and methods. *Educational Review, 8,* 283–286.

Ahn, S. J., Legge, G. E., & Luebker, A. (1995). Printed cards for measuring low-vision reading speed. *Vision Research, 35,* 1939–1944.

Akutsu, H., Bedell, H. E., & Patel, S. S. (2000). Recognition thresholds for letters with simulated dioptric blur. *Optometry and Vision Science, 77,* 524–530.

Akutsu, H., Legge, G. E., Showalter, M., Lindstrom, R. L., Zabel, R. W., & Kirby, V. M. (1992). Contrast sensitivity and reading through multifocal intraocular lenses. *Archives of Ophthalmology, 110,* 1076–1080.

Alexander, K. R., Fishman, G. A., & Derlacki D. J. (1996). Intraocular light scatter in patients with retinitis pigmentosa. *Vision Research, 36,* 3703–3709.

Alexander, K. R., Xie, W., & Derlacki, D. J. (1994). Spatial-frequency characteristics of letter identification. *Journal of the Optical Society of America A, Optics, Image Science, & Vision, 11,* 2375–2382.

American Psychiatric Association. (1994). *Diagnostic and statistical manual of mental disorders* (4th ed.). Washington, DC.

Anstis, S. M. (1974). A chart demonstrating variations in acuity with retinal position. *Vision Research, 14,* 589–592.

Aquilante, K., Yager, D., Morris, R. A., & Khmelnitsky, F. (2001). Low-vision patients with age-related maculopathy read RSVP faster when word duration varies according to word length. *Optometry and Vision Science, 78,* 290–296.

Arditi, A. (1996). Typography, print legibility, and low vision. In R. Cole & B. Rosenthal (Eds.), *Remediation and management of low vision* (pp. 237–248). St. Louis, MO: Mosby.

201

Arditi, A. (1999a). Elicited sequential presentation for low vision reading. *Vision Research, 39,* 4412–4418.

Arditi, A. (1999b). *Making text legible: Designing for people with partial sight.* New York: Lighthouse International. Retrieved November 20, 2004, from http://www.lighthouse.org/print_leg.htm

Arditi, A. (2004). Adjustable typography: An approach to enhancing low vision text accessibility. *Ergonomics, 47,* 469–482.

Arditi, A. (2005). *Color contrast: Designing for people with partial sight and color deficiencies.* Retrieved June 29, 2005, from http://www.lighthouse.org/color contrast.htm

Arditi, A., Cagenello, R., & Jacobs, B. (1995). Effects of aspect ratio and spacing on legibility of small letters. *Investigative Ophthalmology and Visual Science, 36*(Suppl.), 671.

Arditi, A., & Cho, J. (2000a). Do serifs enhance or diminish text legibility? *Investigative Ophthalmology and Visual Science, 41*(Suppl.), 437.

Arditi, A., & Cho, J. (2000b). Letter case and text legibility. *Perception, 29*(Suppl. S), 45.

Arditi, A., & Cho, J. (2005). Serifs and font legibility. *Vision Research, 45,* 2926–2933.

Arditi, A., Knoblauch, K., & Grunwald, I. (1990). Reading with fixed and variable pitch. *Journal of the Optical Society of America A, 7,* 2011–2015.

Arditi, A., Liu, L., & Lynn, W. (1997). Legibility of outline and solid fonts with wide and narrow spacing. In D. Yager (Series Ed.), *Trends in optics and photonics* (Vol. 11, pp. 52–56). Washington, DC: Optical Society of America.

Bailey, I. L., Boyd, L. H., Boyd, W. L., & Clark, M. (1987). Readability of computer display print enlarged for low vision. *American Journal of Optometry and Physiological Optics, 64,* 678–685.

Bailey, I. L., Bullimore, M. A., Greer, R. B., & Mattingly, W. B. (1994). Low vision magnifiers—Their optical parameters and methods for prescribing. *Optometry and Vision Science, 71,* 689–698.

Bailey, I. L., & Lovie, J. E. (1976). New design principles for visual acuity letter charts. *American Journal of Optometry and Physiological Optics, 53,* 740–745.

Baldasare, J., Watson, G. R., Whittaker, S. G., & Miller-Shaffer, H. (1986). The development and evaluation of a reading test for low vision individuals with macular loss. *Journal of Visual Impairment and Blindness, 80,* 7859.

Bartels, P., & Kline, D. W. (2002). Aging effects on the identification of digitally blurred text, scenes and faces: Evidence for optical compensation on everyday tasks in the senescent eye. *Aging International, 27,* 56–72.

Beckmann, P. J., & Legge, G. E. (2002). Preneural limitations on letter identification in central and peripheral vision. *Journal of the Optical Society of America A, Optics, Image Science, & Vision, 19,* 2349–2362.

Beckmann, P. J., Legge, G. E., & Luebker, A. (1991). Reading: Letters, words and their spatial-frequency content. *Society for Information Display 1991 Digest of Technical Papers,* 106–108.

Berger, T. D., Martelli, M., Su, M., Aguayo, M., & Pelli, D. G. (2003). Reading quickly in the periphery. *Vision Sciences Society Abstracts in Journal of Vision, 3,* 806a.

Blakemore, C., & Campbell, F. W. (1969). On the existence of neurones in the human visual system selectively sensitive to the orientation and size of retinal images. *Journal of Physiology, 203,* 237–260.

Boder, E. (1973). Developmental dyslexia: A diagnostic approach based on three atypical reading–spelling patterns. *Developmental Medicine and Child Neurology, 15,* 663–687.

Bouma, H. (1970). Interaction effects in parafoveal letter recognition. *Nature, 226,* 177–178.

Bouma, H. (1973). Visual interference in the parafoveal recognition of initial and final letters of words. *Vision Research, 13,* 767–782.

Bouma, H., & de Voogd, A. H. (1974). On the control of eye saccades in reading. *Vision Research, 14,* 273–284.

Bowers, A. R. (2000). Eye movements and reading with plus-lens magnifiers. *Optometry and Vision Science, 77,* 25–31.

Bowers, A. R., Meek, C., & Stewart, N. (2001). Illumination and reading performance in age-related macular degeneration. *Clinical and Experimental Ophthalmology, 84,* 139–147.

Bowers, A. R., Woods, R. L., & Peli, E. (2004). Preferred retinal locus and reading rate with four dynamic text presentation formats. *Optometry and Vision Science, 81,* 205–213.

Bowers, J. S., Vigliocco, G., & Haan, R. (1998). Orthographic, phonological, and articulatory contributions to masked letter and word priming. *Journal of Experimental Psychology: Human Perception & Performance, 24,* 1705–1719.

Breitmeyer, B. (1980). Unmasking visual masking: A look at the "why" behind the veil of the "how." *Psychological Review, 87,* 52–69.

Brindley, G. S. (1970). *Physiology of the retina and visual pathway* (2nd ed.). Baltimore: Williams & Wilkins.

Brindley, G. S., & Lewin, W. S. (1968). The sensations produced by electrical stimulation of the visual cortex. *Journal of Physiology, 196,* 479–493.

Bringhurst, R. (2004). *The elements of typographic style* (3rd ed.). Vancouver, Canada: Hartley & Marks.

Buchel, C., Price, C., Frackowiak, R. S., & Friston K. (1998). Different activation patterns in the visual cortex of late and congenitally blind subjects. *Brain, 121,* 409–419.

Buettner, M., Krischer, C., & Meissen, R. (1985). Characterization of gliding text as a reading stimulus. *Bulletin of the Psychonomic Society, 23,* 479–482.

Bullimore, M. A., & Bailey, I. (1995). Reading and eye movements in age-related maculopathy. *Optometry and Vision Science, 72,* 125–138.

Buonomano, D. (1998). Cortical plasticity: From synapses to maps. *Annual Review of Neuroscience, 21,* 149–186.

Burr, D. C., Morrone, M. C., & Ross, J. (1994). Selective suppression of the magnocellular visual pathway during saccadic eye movements. *Nature, 371,* 511–513.

Campbell, F. W. (1957). The depth of field of the human eye. *Optica Acta, 4,* 157–164.

Campbell, F. W., & Green, D. G. (1965). Optical and retinal factors affecting visual resolution. *Journal of Physiology, 181,* 576–593.

Campbell, F. W., & Robson, J. G. (1968). Application of fourier analysis to the visibility of gratings. *Journal of Physiology, 197,* 551–566.

Carpenter, R. H. S. (1988). *Movements of the eyes* (2nd ed.). London: Pion.

Carver, R. P. (1976). Word length, prose difficulty and reading rate. *Journal of Reading Behavior, 8,* 193–204.

Carver, R. P. (1982). Optimal rate of reading prose. *Reading Research Quarterly, 18,* 56–88.

Carver, R. P. (1989). What does maximum oral reading rate measure? *Yearbook of the national Reading Conference, 38,* 421–425.

Carver, R. P. (1990). *Reading rate: A review of research and theory.* San Diego, CA: Academic.

Carver, R. P. (2000). *Technical manual for the rate level test (Form A, Form B, and Form C).* Kansas City, MO: Revrac.

Casco, C., Campana, G., Grieco, A., Musetti, S., & Perrone, S. (2003). Hyper-vision in a patient with central and paracentral vision loss reflects cortical reorganization. *Visual Neuroscience, 20,* 501–510.

Castelhano, M. S., & Muter, P. (2001). Optimizing the reading of electronic text using rapid serial visual presentation, *Behaviour and Information Technology, 20,* 237–247.

Castles, A., & Coltheart, M. (1993). Varieties of developmental dyslexia, *Cognition, 47,* 149–180.

Castles, A., Datta, H., Gayan, J., & Olson, R. K. (1999). Varieties of developmental reading disorder: Genetic and environmental influences. *Journal of Experimental Child Psychology, 72,* 73–94.

Cattell, J. M. (1886). The time taken up by cerebral operations. *Mind, 11,* 220–242, 377–392, 524–538.

Cattell, J. M. (1947). On the time required for recognizing and naming letters and words, pictures and colors (R. S. Woodworth, Trans.). *Man of Science, 1,* 13–25. (Original work published 1885)

Charness, N., & Dijkstra, K. (1999). Age, luminance, and print legibility in homes, offices, and public places. *Human Factors, 41,* 173–193.

Chase, C., Ashourzadeh, A., Kelly, C., Monfette, S., & Kinsey, K. (2003). Can the magnocellular pathway read? Evidence from studies of color. *Vision Research, 43,* 1211–1222.

Cheung, S.-H., & Legge, G. E. (2005). Functional and cortical adaptations to central vision loss. *Visual Neuroscience, 22,* 187–201.

Chung, S. T. L. (2005, March). *Two approaches to reduce crowding do not lead to improved reading speed in patients with age-related macular degeneration.* Paper presented at Vision 2005 International Low Vision Conference, London, England.

Chung, S. T. L. (2002). The effect of letter spacing on reading speed in central and peripheral vision. *Investigative Ophthalmology and Visual Science, 43,* 1270–1276.

Chung, S. T. L., & Legge, G. E. (1997). Is reading with a central scotoma like reading with normal peripheral vision? *Optometry and Vision Science, 74*(Suppl.), 153.

Chung, S. T. L., Legge, G. E., & Cheung, S.-H. (2004). Letter recognition and reading speed in peripheral vision benefit from perceptual learning. *Vision Research, 44,* 695–709.

Chung, S. T. L., Legge, G. E., & Tjan, B. S. (2002). Spatial-frequency characteristics of letter identification in central and peripheral vision. *Vision Research, 42,* 2137–2152.

Chung, S. T. L., Levi, D. M., & Legge, G. E. (2001). Spatial-frequency and contrast properties of crowding. *Vision Research, 41,* 1833–1850.

Chung, S. T. L., & Mansfield, J. S. (1999). Does reading mixed-polarity text improve reading speed in peripheral vision? *Vision '99 International Low Vision Conference Abstracts,* 249.

Cohen, L., Dehaene, S., Naccache, L., Lehericy, S., Dehaene-Lambertz, G., Henaff, M., et al. (2000). The visual word form area: Spatial and temporal characterization of an initial stage of reading in normal subjects and posterior split-brain patients. *Brain, 123,* 291–307.

Cohen, L., Lehericy, S., Chochon, F., Lemer, C., Rivaud, S., & Dehaene, S. (2002). Language-specific tuning of visual cortex? Functional properties of the visual word form area. *Brain, 125,* 1054–1069.

Cohen, J. M., & Waiss, B. (1991a). Reading speed in normal observers through different forms of equivalent power low vision devices. *Optometry and Vision Science, 68,* 127–131.

Cohen, J. M., & Waiss, B. (1991b). Reading speed through different equivalent power low vision devices with identical field of view. *Optometry and Vision Science, 68,* 795–797.

Coltheart, M., Curtis, B., Atkins, P., & Halter, M. (1993). Models of reading aloud: Dual-route and parallel-distributed-processing approaches. *Psychological Review, 100,* 589–608.

Coltheart, M., Rastle, K., Perry, C., Langdon, R., & Ziegler, J. (2001). DRC: A dual-route cascaded model of visual word recognition and reading aloud. *Psychological Review, 108,* 204–256.

Cornelissen, P. L., Hansen, P. C., Gilchrist, I., Cormack, F., Essex, J., & Frankish, C. (1998). Coherent motion detection and letter position encoding. *Vision Research, 38,* 2181–2191.

Cornelissen, P. L., Hansen, P. C., Hutton, J. L., Evangelinou, V., & Stein, J. F. (1998). Magnocellular visual function and children's single word reading. *Vision Research, 38,* 471–482.

Culham, L. E., Chabra, A., & Rubin, G. S. (2004). Clinical performance of electronic, head-mounted, low-vision devices. *Ophthalmic and Physiological Optics, 24,* 281–290.

Curcio C. A., Owsley, C., & Jackson, G. R. (2000). Spare the rods save the cones in aging and age-related maculopathy. *Investigative Ophthalmology and Visual Science, 41,* 2015–2018.

Daly, S. (2001). Analysis of subtriad addressing algorithms by visual system models. *Society for Information Display 2001 Digest of Technical Papers,* 1200–1203.

Dehaene, S., Le Clec, H., G., Poline, J. B., Bihan, D. L., & Cohen, L. (2002). The visual word form area: A prelexical representation of visual words in the left fusiform gyrus. *Neuroreport, 13,* 321–325.

Dehaene, S., Naccache, L., Cohen, L., Bihan, D. L., Mangin, J. F., Poline, J. B., et al. (2001). Cerebral mechannisms of word masking and unconscious repetition priming. *Nature Neuroscience, 4,* 752–758.

DeMarco, L., & Massof, R. (1997). Distributions of print sizes in U.S. newspapers. *Journal of Visual Impairment and Blindness, 97,* 9–13.

Demb, J., Boynton, G., & Heeger, D. (1998). Functional magnetic resonance imaging of early visual pathways in dyslexia. *Journal of Neuroscience, 18,* 6939–6951.

Demb, J., Boynton, G., Best, M., & Heeger, D. (1998). Psychophysical evidence for a magnocellular pathway deficit in dyslexia. *Vision Research, 38,* 1555–1559.

Dowdeswell, H. J., Slater, A. M., Broomhall, J., & Tripp, J. (1995). Visual deficits in children born at less than 32 weeks' gestation with and without major ocular pathology and cerebral damage. *British Journal of Ophthalmology, 79,* 447–452.

Eden, G. F., VanMeter, J. W., Rumsey, J. M., Maisog, J. M., Woods, R. P., & Zeffiro, T. A. (1996). Abnormal processing of visual motion in dyslexia revealed by functional brain imaging. *Nature, 382,* 66–69.

Ehrlich, D. (1987). A comparative study in the use of closed-circuit television reading machines and optical aids by patients with retinitis pigmentosa and maculopathy *Ophthalmic and Physiological Optics, 7,* 293–302.

Elbert, T., Pantev, C., Wienbruch, C., Rockstroh, B., & Taub, E. (1995). Increased cortical representation of the fingers of the left hand in string players. *Science, 270,* 305–307.

Eldred, K. B. (1992). Optimal illumination for reading in patients with age-related maculopathy. *Optometry and Vision Science, 69,* 46–50.

Elliott, D. B., Trukolo-Ilic, M., Strong, J. G., Pace, R., Plotkin, A., & Bevers, P. (1997). Demographic characteristics of the vision disabled elderly. *Investigative Ophthalmology and Visual Science, 38,* 2566–2575.

Engbert, R., Longtin, A., & Kliegl, R. (2002). A dynamical model of saccade generation in reading based on spatially distributed lexical processing. *Vision Research, 42,* 621–636.

Estes, W. (1978). Perceptual processing in letter recognition and reading. In E. Carterette & M. Friedman (Eds.), *Handbook of perception, IX* (pp. 163–220). New York: Academic.

Eye Diseases Prevalence Research Group. (2004). Causes and prevalence of visual impairment among adults in the United States. *Archives of Ophthalmology, 122,* 477–485.

Farmer, M., & Klein, R. (1995). The evidence for a temporal processing deficit linked to dyslexia: A review. *Psychonomic Bulletin & Review, 2,* 460–493.

Farrell, G. (1956). *The story of blindness.* Cambridge, MA: Harvard University Press.

Faye, E. E. (1976). *Clinical low-vision.* Boston, MA: Little, Brown.

Ferris, F. L., Kassoff, A., Bresnick, G. H., & Bailey, I. L. (1982). New visual acuity charts for clinical research. *American Journal of Ophthalmology, 94,* 91–96.

Fine, E. M. (2001). Does meaning matter? The impact of word knowledge on lateral masking. *Optometry and Vision Science, 78,* 831–838.

Fine, E. M., Hazel, C. A., Petre, K. L., & Rubin, G. S. (1999). Are the benefits of sentence context different in central and peripheral vision? *Optometry and Vision Science, 76,* 764–769.

Fine, E. M., Kirschen, M. P., & Peli, E. (1996). The necessary field of view to read with an optimal stand magnifier. *Journal of the American Optometric Association, 67,* 382–389.

Fine, E. M., & Peli, E. (1996). The role of context in reading with central field loss. *Optometry and Vision Science, 73,* 533–539.

Fine, E. M., Peli, E., & Reeves, A. (1997). Simulated cataract does not reduce the benefit of RSVP. *Vision Research, 37,* 2639–2647.

Fine, E. M., & Rubin, G. (1999). Reading with simulated scotomas: Attending to the right is better than attending to the left. *Vision Research, 39,* 1039–1048.

Fine, I., & Jacobs, R. A. (2002). Comparing perceptual learning across tasks: A review. *Journal of Vision, 2,* 190–203.

Fletcher, D. C., Schuchard, R. A., Livingstone, C. L., Crane, W. G., & Hu, S. Y. (1994). Scanning laser ophthalmoscope macular perimetry and applications for low vision rehabilitation clinicians. *Low Vision and Vision Rehabilitation, 7,* 257–265.

Fletcher, D. C., Schuchard, R. A., & Watson, G. (1999). Relative locations of macular scotomas near the PRL: Effect on low vision reading. *Journal of Rehabilitation Research and Development, 36,* 356–364.

Forbes, T. W., & Holmes, R. S. (1939). Legibility distances of highway destination signs in relation to letter height, width, and reflectorization. *Proceedings of the Highway Research Board U.S.A., 19,* 321–335.

Forbes, T. W., Moskowitz, K., & Morgan, G. (1950). A Comparison of lower-case and capital letters for highway signs. *Proceedings of the 30th annual meeting of the Highway Research Board* (pp. 355–373). Washington, DC: Highway Research Board.

Forster, K. I. (1970). Visual perception of rapidly presented word sequences of varying complexity. *Perception and Psychophysics, 8,* 215–221.

Foulke, E. (1982). Reading braille. In W. Shiff & E. Foulke (Eds.), *Tactual perception: A source book* (pp. 168–208). Cambridge, England: Cambridge University Press.

Garvey, P., Pietrucha, M., & Meeker, D. (1998). Clearer road signs ahead. *Ergonomics in Design, 6,* 7–11.

Geisler, W. S. (1989). Sequential ideal-observer analysis of visual discriminations. *Psychological Review, 96,* 267–314.

Gelb, I. J. (1963). *A study of writing* (2nd ed.). Chicago, University of Chicago Press.

Gibson, E. J., & Levin, H. (1975). *The psychology of reading.* Cambridge, MA: MIT Press.

Gilchrist, I., Brown, V., & Findlay, J. (1997). Saccades without eye movements. *Nature, 390,* 130–131.

Gill, J. M., & Perera, S. (2003, July). *The tiresias family of fonts.* Retrieved November 20, 2004, from http://www.tiresias.org/reports/marburg.htm

Ginsburg, A. P. (1978). *Visual information processing based on spatial filters constrained by biological data. Vols. I and II.* Unpublished doctoral thesis, Cambridge University, England.

Gould, J., Alfaro, L., Finn, R., Haupt, B., & Minuto, A. (1987). Reading from CRT displays can be as fast as reading from paper. *Human Factors, 29,* 497–517.

Habib, M. (2000). The neurological basis of developmental dyslexia: An overview and working hypothesis. *Brain, 123,* 2373–2399.

Harmon, L. D., & Julesz, B. (1973). Masking in visual recognition: Effects of two-dimensional filtered noise. *Science, 189,* 1194–1197.

Hartley, J. T., Stojack, C. C., Mushaney, T. T., Kiku Annon, T. A., & Lee, D. W. (1994). Reading speed and prose memory in older and younger adults. *Psychology & Aging, 9,* 216–23.

Hayes, J. S., Yin, V. T., Piyathaisere, D, Weiland, J. D., Humayun, M. S., & Dagnelie, G. (2003). Visually guided performance of simple tasks using simulated prosthetic vision. *Artificial Organs, 27,* 1016–28.

Heinen, S. J., & Skavenski, A. A. (1991). Recovery of visual responses in foveal V1 neurons following bilateral foveal lesions in adult monkey. *Experimental Brain Research, 83,* 670–674.

Hensil, J., & Whittaker, S. G. (2000). Visual reading vs. auditory reading? Sighted persons and persons with low vision. *Journal of Visual Impairment and Blindness, 94,* 762–770.

Hering, E. (1879). Ober muskelgerausche das auges. Sitzungsberichte der Akademie der Wissenschaften in Wien. *Mathematisch-Naturwissenschaftliche Klasse, Abt. 11179,* 137–154.

Herse, P., & Bedell, H. (1989). Contrast sensitivity for letter and grating targets under various stimulus conditions. *Optometry and Vision Science, 66,* 774–781.

Huey, E. B. (1968). *The psychology and pedagogy of reading.* Cambridge, MA: MIT Press. (Original work published 1908)

Humayun, M. S., & de Juan, E. (1998). Artificial vision. *Eye, 12,* 605–607.

Humayun, M. S., de Juan, E., Weiland, J. D., Dagnelie, G., Katona, S., Greenberg, R., et al. (2003). Visual perception in a blind subject with a chronic microelectronic retinal prosthesis. *Vision Research, 43,* 2573–2581.

IESNA, Illuminating Engineering Society of North America. (2000). *Lighting handbook: Reference & application* (9th ed.). New York: Illuminating Engineering Society of North America.

Irlen, H. (1991). *Reading by the colours: Overcoming dyslexia and other reading disabilities through the Irlen method.* New York: Avery.

Jackson, M., & McClelland, J. (1975). Sensory and cognitive determinants of reading speed. *Journal of Verbal Learning and Verbal Behavior, 14,* 565–574.

Jackson, M., & McClelland, J. (1979). Process determinants of reading speed. *Journal of Experimental Psychology: General, 108,* 151–181.

Jarvis, S. H., & Chung, S. T. L. (2004). Effect of optical blur on reading speed. *Optometry and Vision Science, 81*(Suppl.), 138.

Javal, L. E, (1879). Essai sur la physiologie de la lecture. *Annales d'Oculistique, 82,* 242–253.

Jorna, G. C., & Snyder, H. L. (1991). Image quality determines differences in reading performance and perceived image quality with CRT and hard-copy displays. *Human Factors, 33,* 459–469.

Juola, J. F., Tiritoglu, A., & Pleunis, J. (1995). Reading text presented on a small display. *Applied Ergonomics, 26,* 227–229.

Juola, J. F., Ward, N. J., & McNamara, T. (1982). Visual search and reading of rapid serial presentations of letter strings, words, and text. *Journal of Experimental Psychology: General, 111*, 208–227.

Just, M. A., & Carpenter, P. A. (1980). A theory of reading: From eye fixations to comprehension. *Psychological Review, 87*, 329–354.

Just, M. A., & Carpenter, P. A. (1987). *The psychology of reading and language comprehension.* Boston, MA: Allyn & Bacon.

Kallie, C. S., Cheung, S. H., Legge, G. E., Owsley, C., & McGwin, G. (2005). Nonlinear mixed effects modeling as an estimation procedure for sparse MNREAD data. *Investigative Ophthalmology and Visual Science, 46*, 4589.

Kauffman, T., Theoret, H., & Pascual-Leone, A. (2002). Braille character discrimination in blindfolded human subjects. *Neuroreport, 13*, 571–574.

Kilgarriff, A. (1997). Putting frequencies in the dictionary. *International Journal of Lexicography, 10*, 135–155.

Kitchel, J. E. (2004, March). *Large print: Guidelines for optimal readability and APHontTM a font for low vision.* Louisville, KY: The American Printing House for the Blind, Inc. Retrieved November 20, 2004, from http://www.aph.org/edresearch/lpguide.htm

Kline, D. W., Buck, K., Sell, Y., Bolan, T. L., & Dewar, R. E. (1999). Older observers' tolerance of optical blur: Age differences in the identification of defocused text signs. *Human Factors, 41*, 356–364.

Klitz, T. S., Mansfield, J. S., & Legge, G. E. (1995). Reading speed is affected by font transitions. *Investigative Ophthalmology and Visual Science, 36*(Suppl. S), 670.

Knoblauch, K., Arditi, A., & Szlyk, J. (1991). Effects of chromatic and luminance contrast on reading. *Journal of the Optical Society of America A, 8*, 428–439.

Knuth, D. E. (1986). *The METAFONT book.* Reading, MA: Addison-Wesley.

Kooi, F. L., Toet, A., Tripathy, S. P., & Levi, D. M. (1994). The effect of similarity and duration on spatial interaction in peripheral vision. *Spatial Vision, 8*, 255–279.

Kowler, E., & Anton, S. (1987). Reading twisted text: Implications for the role of saccades. *Vision Research, 27*, 45–60.

Kučera, H., & Francis, W. N. (1967). *Computational analysis of present day American English.* Providence, RI: Brown University Press.

Kujala, T., Alho, K., & Naatanen, R. (2000). Cross-modal reorganization of human cortical functions. *Trends in Neuroscience, 23*, 115–120.

LaGrow, S. J. (1986). Assessing optimal illumination for visual response accuracy in visually impaired adults. *Journal of Visual Impairment and Blindness, 83*, 888–895.

Lamare, M. (1892). Des mouvements des yeux dans la lecture. *Bulletin et Memoire de la Societe Frangaise d'Ophthalmologie, 10*, 354–364.

Latham, K., & Whitaker, D. (1996). A comparison of word recognition and reading performance in foveal and peripheral vision. *Vision Research, 36*, 2665–2674.

Lawson, A. S., & Agner D. (1990). *Printing types: An introduction.* Boston, MA: Beacon.

Leat S. J., Legge, G. E., & Bullimore, M. (1999). What is low vision? A re-evaluation of definitions. *Optometry and Vision Science, 76*, 198–211.

Lee, H.-W., Gefroh, J., Legge, G. E., & Kwon, M. Y. (2003, November). *Training improves reading speed in peripheral vision: Is it due to attention?* Poster session presented at the annual meeting of the Psychonomic Society, Vancouver, BC, Canada.

Legge, G. E. (1976). Adaptation to a spatial impulse: Implications for Fourier transform models of visual processing. *Vision Research, 16*, 1407–1418.

Legge, G. E. (1978). Sustained and transient mechanisms in human vision: Temporal and spatial properties. *Vision Research, 18,* 69–81.

Legge, G. E., & Foley, J. M. (1980). Contrast masking in human vision. *Journal of the Optical Society of America, 70,* 1458–1471.

Legge, G. E., Hooven, T. A., Klitz, T. S., Mansfield, J. S., & Tjan, B. S. (2002). Mr. Chips 2002: New insights from an ideal-observer model of reading. *Vision Research, 42,* 2219–2234.

Legge, G. E., Klitz, T. S., & Tjan, B. S. (1997). Mr. Chips: An ideal-observer model of reading. *Psychological Review, 104,* 524–553.

Legge, G. E., Lee, H.-W., Owens, D., Cheung, S.-H., & Chung, S. T. L. (2002). Visual span: A sensory bottleneck on reading speed [Abstract]. *Journal of Vision, 2,* 279a.

Legge, G. E., Madison, C., & Mansfield, J. S. (1999). Measuring braille reading speed with the MNREAD test. *Visual Impairment Research, 1,* 131–145.

Legge, G. E., Mullen, K. T., Woo, G. C., & Campbell, F. W. (1987). Tolerance to visual defocus. *Journal of the Optical Society of America, 4,* 851–863.

Legge, G. E., Rubin, G. S., & Schleske, M. M. (1986). Contrast-polarity effects in low-vision reading. In G. C. Woo (Ed.), *Low vision: Principles and applications* (pp. 288–307). New York: Springer-Verlag.

Lei, H., & Schuchard, R. A. (1997). Using two preferred retinal loci for different lighting conditions in patients with central scotomas. *Investigative Ophthalmology and Visual Science, 38,* 1812–1818.

Levi, D. M., Klein, S. A., & Aitsebaomo, A. P. (1985). Vernier acuity, crowding and cortical magnification. *Vision Research, 25,* 963–77.

Li, L., Nugent, A. K., & Peli, E. (2000). Recognition of jagged (pixelated) letters in the periphery. *Visual Impairment Research, 2,* 143–154.

Livingstone, M. S., Rosen, G. D., Drislane, F. W., & Galaburda, A. M. (1991). Physiological and anatomical evidence for a magnocellular defect in developmental dyslexia. *Proceedings of the National Academy of Sciences USA, 88,* 7943–7947.

Loomis, J. (1981). On the tangibility of letters and Braille. *Perception & Psychophysics, 29,* 37–46.

Lott, L. A., Schneck, M. E., Haegerstrom-Portnoy, G., Brabyn, J. A., Gildengorin, G. L., & West, C. G. (2001). Reading performance in older adults with good acuity. *Optometry and Vision Science, 78,* 316–324.

Lovegrove, W. J., Bowling, A., Badcock, D., & Blackwood, M. (1980). Specific reading disability: Differences in contrast sensitivity as a function of spatial frequency, *Science, 210,* 439–440.

Lovie-Kitchin, J. E., & Bailey, I. L. (1981). Task complexity and visual acuity in senile macular degeneration. *Australian Journal of Optometry, 64,* 235–242.

Lovie-Kitchin, J. E., Bowers, A., & Woods, R. (2000). Oral and silent reading performance with macular degeneration. *Ophthalmology and Physiological Optics, 20,* 360–370.

Lovie-Kitchin, J. E., & Woo, G. C. (1988). Effect of magnification and field of view on reading speed using a CCTV. In G. C. Woo (Ed.), *Low vision: Principles and applications* (pp. 308–322). New York: Springer-Verlag.

Lowe, J. B., & Drasdo, N. (1990). Efficiency in reading with closed-circuit television for low vision. *Ophthalmic and Physiological Optics, 10,* 225–233.

Ludvigh, E. (1941). Extrafoveal visual acuity as measured with Snellen test-letters. *American Journal of Ophthalmology, 24,* 303–310.

Lund, R., & Watson, G. (1997). *The CCTV book.* Froland, Norway: Synsforum.

Majaj, N. J., Liang, Y. X., Martelli, M., Berger, T. D., & Pelli, D. G. (2003). Channel for reading. *Vision Sciences Society Abstracts in Journal of Vision, 3,* 813a.

Majaj, N. J., Pelli, D. G., Kurshan, P., & Palomares, M. (2002). The role of spatial frequency channels in letter identification. *Vision Research, 42,* 1165–1184.

Mancil, G. L., & Nowakowski, R. (1986). Evaluation of reading speed with four low vision aids. *American Journal of Optometry and Physiological Optics, 63,* 708–713.

Mansfield, J. S., Ahn, S. J., Legge, G. E., & Luebker, A. (1993). A new reading-acuity chart for normal and low vision. *Ophthalmic and Visual Optics/Noninvasive Assessment of the Visual System Technical Digest, 3,* 232–235.

McClelland, J. L., & Rumelhart, D. E. (1981). An interactive activation model of context effects in letter perception: Part 1. An account of basic findings. *Psychological Review, 88,* 375–407.

McConkie, G. W., Kerr, P. W., Reddix, M. D., & Zola, D. (1988). Eye movement control during reading: I. The location of initial eye fixations on words. *Vision Research, 28,* 1107–1118.

McConkie, G. W., Kerr, P. W., Reddix, M. D., Zola, D., & Jacobs, A. M. (1989). Eye movement control during reading: II. Frequency of refixating a word. *Perception & Psychophysics, 46,* 245–253.

McConkie, G. W, & Rayner, K. (1975). The span of the effective stimulus during a fixation in reading. *Perception & Psychophysics, 17,* 578–586.

Mellor, C. M. (2006). *Louis Braille: A touch of genius.* Boston, MA: National Braille Press.

Merabet, L. B., Rizzo, J. F., Amedi, A., Somers, D. C., & Pascual-Leone, A. (2005). What blindness can tell us about seeing again: merging neuroplasticity and neuroprostheses. *Nature Reviews Neuroscience, 6,* 71–77.

Merzenich, M. M., Nelson, R. J., Stryker, M. P., Cynader, M. S., Schoppmann, A., & Zook, J. M. (1984). Somatosensory cortical map changes following digit amputation in adult monkeys. *Journal of Comparative Neurology, 224,* 591–605.

Mewhort, D. J. K., Campbell, A. J., Marchetti, F. M., & Campbell, J. I. D. (1981). Identification, localization, and iconic memory: An evaluation of the bar-probe task. *Memory & Cognition, 9,* 50–67.

Morris, R. A., Aquilante, K., Yager, D., & Bigelow, C. (2002). *Serifs slow RSVP reading at very small sizes but don't matter at larger sizes.* Society for Information Display 2002 Digest of Technical Papers, 244–247.

Morris, R. A., Hersch, R. D., & Coimbra, A. (1998, March 30–April 3). *Legibility of condensed perceptually-tuned grayscale fonts.* Paper presented at the Electronic Publishing, Artistic Imaging and Digital Typography, Proc. EP '98 and RIDT '98 Conferences, LNCS 1375, St. Malo: England.

The National Library Service. (2000). *Braille into the next millennium.* Washington, DC: The Library of Congress.

National Research Council. (1980). Recommended standards for the clinical measurement and specification of visual acuity. *Advances in Ophthalmology, 41,* 103–148.

National Research Council. (2002). V*isual impairments: Determining eligibility for social security benefits.* Washington DC: National Academy Press.

Nazir, T. A., O'Regan, J. K., & Jacobs, A. M. (1991). On words and their letters. *Bulletin of the Psychonomic Society, 29,* 171–174.

Nilsson, U., Frennesson, C., & Nilsson, S. (1998). Location and stability of a newly established eccentric retinal locus suitable for reading, achieved through training of patients with a dense central scotoma. *Optometry and Vision Science, 75,* 873–878.

Nolan, C., & Kederis, C. (1969). *Perceptual factors in braille word recognition* (Research Series No. 20). New York: American Foundation for the Blind.

Normann, R. A, Maynard, E. M., Rousche, P. J., & Warren, D. J. (1999). A neural interface for a cortical vision prosthesis. *Vision Research, 39,* 2577–2587.

O'Brien, B. A., Mansfield, J. S., & Legge, G. E. (2000). The effect of contrast on reading speed in dyslexia. *Vision Research, 40,* 1921–1935.

O'Regan, J. K. (1990). Eye movements and reading. In E. Kowler (Ed.), *Eye movements and their role in visual and cognitive processes* (pp. 395–453). New York: Elsevier Science.

O'Regan, J. K. (1991). Understanding visual search and reading using the concept of stimulus "grain." *IPO Annual Progress Report, 26,* 96–108.

O'Regan, J. K., Bismuth, N., Hersch, R. D., & Pappas, A. (1996). *Legibility of perceptually-tuned grayscale fonts.* Proceedings of the IEEE International Conference on Image Processing, ICIP '96 (P. Delogne, Ed.), Vol. 1, 537–540.

O'Regan, J. K., Levy-Schoen, A., & Jacobs, A. M. (1983). The effect of visibility on eye-movement parameters in reading. *Perception & Psychophysics, 34,* 457–464.

Ortiz, A. (2002). *Perceptual properties of letter reading in central and peripheral vision.* Unpublished doctoral dissertation, University of Minnesota, Minneapolis.

Ortiz, A., Chung, S. T. L., Legge, G. E., & Jobling, J. T. (1999). Reading with a head-mounted video magnifier. *Optometry and Vision Science, 76,* 755–763.

Owsley, C., & McGwin, G. (1999). Vision impairment and driving. *Survey of Ophthalmology, 43,* 535–550.

Pammer, K., Hansen, P. C., Kringelbach, M. L., Holliday, I., Barnes, G., Hillebrand, A., et al. (2004). Visual word recognition: The first half second. *Neuroimage, 22,* 1819–1825.

Parish, D. H., & Sperling, G. (1991). Object spatial frequencies, retinal spatial frequencies, and the efficiency of letter discrimination. *Vision Research, 31,* 1399–1415.

Pascual-Leone, A., & Hamilton, R. (2001). The metamodal organization of the brain. *Progress in Brain Research, 134,* 427–45.

Pascual-Leone, A., & Torres, F. (1993). Plasticity of the sensorimotor cortex representation of the reading finger in Braille readers. *Brain, 116,* 39–52.

Pastoor, S. (1990). Legibility and subjective preference for color combinations in text. *Human Factors, 32,* 157–171.

Paulesu, E., McCrory, E., Fazio, F., Menoncello, L., Brunswick, N., Cappa, S. F., et al. (2000). A cultural effect on brain function. *Nature Neuroscience, 3,* 91–96.

Peli, E. (1986). Control of eye movement with peripheral vision: Implications for training of eccentric viewing. *American Journal of Optometry and Physiological Optics, 63,* 113–118.

Peli, E., & Siegmund, W. P. (1995). Fiber-optic reading magnifiers for the visually impaired. *Journal of the Optical Society of America, A, 12,* 2274–2286.

Pelli, D. G., Burns, C., Farell, B., & Moore-Page, D. C. (in press). *Feature detection and letter identification.* Vision Research.

Pelli, D. G., Farell, B., & Moore, D. C. (2003). The remarkable inefficiency of word recognition. *Nature, 423,* 752–756.

Pelli, D. G., Palomares, M., & Majaj, N. (2004). Crowding is unlike ordinary masking: Distinguishing feature integration from detection. *Journal of Vision, 4,* 1136–1169.

Perera, S. (2001). *LPfont: An investigation into the legibility of large print typefaces.* Retrieved November 20, 2004, from http://www.tiresias.org/fonts/lpfont/report/index.htm

Pokorny, J., Graham, C. H., & Lanson, R. N. (1968). Effect of wavelength on foveal grating acuity. *Journal of the Optical Society of America A, 58,* 1410–1414.

Price, C., & Devlin, C., (2003). The myth of the visual word form area. *Neuroimage, 19,* 463–481.

Rahman, T., & Muter, P., (1999). Designing an interface to optimize reading with small display windows. *Human Factors, 41,* 106–117.

Rayner, K. (1986). Eye movements and the perceptual span in beginning and skilled readers. *Journal of Experimental Child Psychology, 41,* 211–236.

Rayner, K. (1998). Eye movements in reading and information processing: 20 years of research. *Psychological Bulletin, 124,* 372–422.

Rayner, K., & Bertera, J. H. (1979). Reading without a fovea. *Science, 206,* 468–469.

Rayner, K., & Fischer, M. (1996). Mindless reading revisited: Eye movements during reading and scanning are different. *Perception & Psychophysics, 58,* 734–747.

Rayner, K., & McConkie, G. (1976). What guides a reader's eye movements? *Vision Research, 16,* 829–837.

Rayner, K., McConkie, G. W., & Zola, D. (1980). Integrating information across eye movements. *Cognitive Psychology, 12,* 206–226.

Rayner, K., & Pollatsek, A. (1989). *The psychology of reading.* Englewood Cliffs, NJ: Prentice-Hall.

Rayner, K, Sereno, S., & Raney, G. (1996). Eye movement control in reading: A comparison of two types of models. *Journal of Experimental Psychology: Human Perception and Performance, 22,* 1188–1200.

Rayner, K., Well, A. D., & Pollatsek, A. (1980). Asymmetry of the effective visual field in reading. *Perception & Psychophysics, 27,* 537–544.

Reicher, G. M. (1969). Perceptual recognition as a function of the meaningfulness of stimulus material. *Journal of Experimental Psychology, 81,* 275–280.

Reichle, E. D., Pollatsek, A., Fisher, D. L., & Rayner, K. (1998). Toward a model of eye movement control in reading. *Psychological Review, 105,* 125–157.

Robinson, G. L., & Foreman, P. J. (1999). Scotopic sensitivity/Irlen syndrome and the use of coloured filters: A long-term placebo controlled and masked study of reading achievement and perception of ability. *Perception and Motor Skills, 89,* 83–113.

Roufs, J., & Boschmann, M. C. (1997). Text quality metrics for visual display units: I. Methodological aspects. *Display, 18,* 37–43.

Rubin, G. S., & Turano, K. (1992). Reading without saccadic eye movements. *Vision Research, 32,* 895–902.

Rubin, G. S., & Turano, K. (1994). Low vision reading with sequential word presentation. *Vision Research, 34,* 1723–1733.

Rumney, N. J., & Leat, S. J. (1994). Why do low vision patients still read slowly with a low vision aid? In A. C. Kooijman, P. L. Looijestijn, J. A. Welling, & G. J. van der Wildt (Eds.), *Low vision—Research and new developments to rehabilitation* (pp. 269–274). Amsterdam, The Netherlands: IOS Press.

Sadato, N., Pascual-Leone A., Grafman, J., Ibanez, V., Deiber, M. P., Dold, G., et al. (1996). Activation of the primary visual cortex by braille reading in blind subjects. *Nature, 380,* 526–528.

Sass, S. M., Legge, G. E., & Lee, H.-W. (2006). Low-vision reading speed: Influences of linguistic inference and aging. *Optometry and Vision Science, 83,* 166–167.

Sathian, K. (1998). Perceptual learning. *Current Science, 75,* 451–457.

Schade, O. H. (1956). Optical and photoelectric analog of the eye. *Journal of the Optical Society of America, 46,* 721–739.

Schaeffel, F., Weiss, S., & Seidel, J. (1999). How good is the match between the plane of the text and the plane of focus during reading? *Ophthalmic and Physiological Optics, 19,* 180–192.

Scharff, L. F., Hill, A. L., & Ahumada, A. J., Jr. (2000). Discriminability measures for predicting readability of text on textured backgrounds. *Optics Express, 6,* 81–91.

Schuchard, R. A., Naseer, S., & de Castro, K. (1999). Characteristics of AMD patients with low vision receiving visual rehabilitation. *Journal of Rehabilitation Research and Development, 36,* 294–302.

Shapley, R., & Enroth-Cugell, C. (1984). Visual adaptation and retinal gain controls. In N. Osborne & G. Chader (Eds.), *Progress in retinal research* (Vol. 3, pp. 263–346). Oxford, England: Pergamon.

Shaywitz, S. E. (1998). Current concepts: Dyslexia. *New England Journal of Medicine, 338,* 308–312.

Shaywitz, S. E., Escobar, M. D., Shaywitz, B. A., Fletcher, J. M., & Makuch, R. (1992). Evidence that dyslexia may represent the lower tail of a normal distribution of reading ability. *New England Journal of Medicine, 326,* 192–193.

Sheedy, J. E., Hayes, J., & Engle, J. (2003). Is all asthenopia the same? *Optometry and Vision Science, 80,* 732–739.

Sheedy, J. E., & McCarthy, M. (1994). Reading performance and visual comfort with scale to gray compared with black-and-white scanned print. *Displays, 15,* 27–30.

Skottun, B. C. (2000a). On the conflicting support for the magnocellular-deficit theory of dyslexia. *Trends in Cognitive Sciences, 4,* 211–212.

Skottun, B. C. (2000b). The magnocellular deficit theory of dyslexia: The evidence from contrast sensitivity. *Vision Research, 40,* 111–127.

Sloan, L. L. (1977). *Reading aids for the partially sighted: A systematic classification and procedure for prescribing.* Baltimore, MD: Williams & Wilkins.

Smith, E. L., III. (1998). Spectacle lenses and emmetropization: The role of optical defocus in regulating ocular development. *Optometry and Vision Science, 75,* 388–398.

Smith, E. L., III, Kee, C. S., Ramamirtham, R., Qiao-Grider, Y., & Hung, L. F. (2005). Peripheral vision can influence eye growth and refractive development in infant monkeys. *Investigative Ophthalmology and Visual Science, 46,* 3965–3972.

Smith, F. (1969). Familiarity of configuration vs. discriminability of features in the visual identification of words. *Psychonomic Science, 14,* 261–263.

Solomon, A. (2004 July 10). The closing of the American book. *New York Times,* p. 17.

Solomon, J. A., & Pelli, D. G. (1994). The visual filter mediating letter identification. *Nature, 369,* 395–397.

Sommerhalder, J., Queghlani, E., Bagnoud, M., Leonards, U., Safran, A. B., & Pelizzone, M. (2003). Simulation of artificial vision: Eccentric reading of isolated words, and perceptual learning. *Vision Research, 43,* 269–283.

Sommerhalder, J., Rappaz, B., & de Haller, R. (2004). Simulation of artificial vision: II. Eccentric reading of full-page text and the learning of this task. *Vision Research, 44,* 1693–1706.

Stanovich, K. E. (1980). Toward an interactive-compensatory model of individual differences in the development of reading fluency. *Reading Research Quarterly, 16,* 32–71.

Stanovich, K. (1999). The sociopsychometrics of learning disabilities. *Journal of Learning Disabilities, 32,* 350–361.

Stanovich, K. (2000). *Progress in understanding reading.* New York: Guilford.

Stanovich, K. E., & West, R. E. (1983). On priming by a sentence context, *Journal of Experimental Psychology: General, 112,* 1–36.

Starr, M. S., & Rayner, K. (2001). Eye movements during reading: Some current controversies. *Trends in Cognitive Science, 5,* 156–163.

Stein, J., Talcott, J., & Walsh, V. (2000). Controversy about the visual magnocellular deficit in developmental dyslexics. *Trends in Cognitive Sciences, 4,* 209–211.

Stein, J. & Walsh, V. (1997). To see but not to read: The magnocellular theory of dyslexia. *Trends in Neuroscience, 20,* 147–152.

Sunness, J. S., Applegate, C. A., Haselwood, D., & Rubin, G. S. (1996). Fixation patterns and reading rates in eyes with central scotomas from advanced atrophic age-related macular degeneration and Stargardt disease. *Ophthalmology, 103,* 1458–1466.

Sunness, J. S., Gonzalez-Baron, J., Applegate, C. A., Bressler, N. M., Tian, Y., Hawkins, B., et al. (1999). Enlargement of atrophy and visual acuity loss in the geographic atrophy form of age-related macular degeneration. *Ophthalmology, 106,* 1768–1779.

Talcott, J. B., Witton, C., Hebb, G., Stoodley, C., Westwood, E., France, S., et al. (2002). On the relationship between dynamic visual and auditory processing and literacy skills: Results from a large primary-school study. *Dyslexia, 8,* 204–225.

Tallal, P. (1984). Temporal or phonetic processing deficit in dyslexia? That is the question. *Applied psycholinguistics, 5,* 167–169.

Tallal, P., & Curtiss, S. (1990). Neurological basis of developmental language disorders. In A. Rothenberger (Ed.), *Brain and behavior in child psychiatry* (pp. 205–216). New York: Springer-Verlag.

Tamaki, C., Kallie, C. S., Legge, G. E., Salomão, S. R., Cudeck, R., & Mansfield, J. S. (2004). Validation of the MNREAD–Portuguese Continuous–Text Reading–Acuity Chart. *Investigative Ophthalmology and Visual Science, 45,* 4358.

Tatler, B., & Wade, N. (2003). On nystagmus, saccades and fixations. *Perception, 32,* 167–184.

Taylor, S. E. (1965). Eye movements in reading: Facts and fallacies. *American Educational Research Journal, 2,* 187–202.

Thompson, R. W., Jr., Barnett, G. D., Humayun, M. S., & Dagnelie, G. (2003). Facial vision using simulated prosthetic pixelized vision. *Investigative Ophthalmology and Visual Science, 34,* 5035–5042.

Thorn, F., & Schwartz, F. (1990). Effects of dioptric blur on Snellen and grating acuity. *Optometry and Vision Science, 67,* 3–7.

Tinker, M. A. (1963). *Legibility of print.* Ames: Iowa State University Press.

Tjan, B. S., Braje, W. L., Legge, G. E., & Kersten, D. (1995). Human efficiency for recognizing 3-D objects in luminance noise. *Vision Research, 35,* 3053–3069.

Trauzettel-Klosinski, S., Teschner, C., Tornow, R.-P., & Zrenner, E. (1994). Reading strategies in normal subjects and in patients with macular scotoma assessed by two new methods of registration. *Neuro-ophthalmology, 14,* 15–30.

Travis, D. S., Bowles, S., Seton, J., & Peppe, R. (1990). Reading from color displays: A psychophysical model. *Human Factors, 32,* 147–156.

Troilo, D., Gottlieb, M. D., & Wallman, J. (1987). Visual deprivation causes myopia in chicks with optic nerve section. *Current Eye Research, 6,* 993–999.

Tyrrell, R. A., Pasquale, T. B., Aten, T., & Francis, E. L. (2001). Empirical evaluation of user responses to reading text rendered using ClearType technologies. *Society for Information Display 2001 Digest of Technical Papers,* 1205–1207.

Virgili, G., Cordaro, C., Bigoni, A., Crovato, S., Cecchini, P., & Menchini, U. (2004). Reading acuity in children: Evaluation and reliability using MNREAD charts. *Investigative Ophthalmology and Visual Science, 45,* 3349–3354.

Virsu, V., & Rovamo, J. (1979). Visual resolution, contrast sensitivity, and the cortical magnification factor. *Experimental Brain Research, 37,* 475–494.

Vitu, F., O'Regan, J. K., Inhoff, A., & Topolski, R. (1995). Mindless reading: Eye-movement characteristics are similar in scanning letter strings and reading texts. *Perception & Psychophysics, 57,* 352–364.

Wade, N., Tatler, B., & Heller, D. (2003). Dodge-ing the issue: Dodge, Javal, Hering and the measurement of saccades in eye-movement research. *Perception, 32,* 793–804.

Walker, P. (1987). Word shape as a cue to the identity of a word: Analysis of the Kučera and Francis (1967) word list. *Quarterly Journal of Experimental Psychology A, 39,* 675–700.

Wallman, J., Gottlieb, M. D., Rajaram, V., & Fugate-Wentzek, L. A. (1987). Local retinal regions control local eye growth and myopia. *Science, 237,* 73–77.

Wallman, J., & Winawer, J. (2004). Homeostasis of eye growth and the question of myopia. *Neuron, 43,* 447–468.

Wandell, B. A. (1995). *Foundations of vision.* Sunderland, MA: Sinauer Associates, Inc.

Warde, B. (1955). The crystal goblet or printing should be invisible. In H. Jacob (Ed.), *The crystal goblet: Sixteen essays on typography* (pp. 11–17). London: Sylvan.

Warrington, E. K., & Shallice, T. (1980). Word-form dyslexia. *Brain, 103,* 99–112.

Watson, G. R., Wright, V., Long, S., & De l'Aune, W. (1996). A low vision reading comprehension test. *Journal of Visual Impairment and Blindness, 90,* 486–494.

Weale, R. A. (1963). *The aging eye.* London: Lewis.

Webster, M. A., Georgeson, M. A., & Webster, S. M. (2002). Neural adjustments to image blur. *Nature Neuroscience, 5,* 839–840.

Westheimer, G. (2003). Visual acuity with reversed-contrast charts. I. Theoretical and psychophysical investigations. *Optometry and Vision Science, 80,* 745–748.

Westheimer, G., Chu, P., Huang, W., Tran, T., & Dister, R. (2003). Visual acuity with reversed-contrast charts. II. Clinical investigation. *Optometry and Vision Science, 80,* 749–752.

Weymouth, F. W. (1958). Visual sensory units and the minimal angle of resolution. *American Journal of Ophthalmology, 46,* 102–113.

Wheeler, D. D. (1970). Processes in word recognition. *Cognitive Psychology, 1,* 59–85.

White, J. M., & Bedell, H. E. (1990). The oculomotor reference in humans with bilateral macular disease. *Investigative Ophthalmology and Visual Science, 31,* 1149–1161.

Whittaker, S. G., Cummings, R. W., & Swieson, L. R. (1991). Saccade control without a fovea. *Vision Research, 31,* 2209–2218.

Whittaker, S. G., & Lovie-Kitchin, J. (1993). Visual requirements for reading. *Optometry and Vision Science, 70,* 54–65.

Wilkins, A. J. (2002). Coloured overlays and their effects on reading speed: A review. *Ophthalmic and Physiological Optics, 22,* 448–454.

Wilkins, A. J. (1995). *Visual stress.* Oxford, England: Oxford University Press.

Wilkins, A. J. (1994). Overlays for classroom and optometric use. *Ophthalmic and Physiological Optics, 14,* 97–99.

Wilkins, A. J., Lewis, E., Smith, F., Rowland, E., & Tweedie, W. (2001). Coloured overlays and their benefit for reading. *Journal of Research in Reading, 24,* 41–64.

Williams, M. C., May, J. G., Solman, R., & Zhou, H. (1995). The effects of spatial filtering and contrast reduction on visual search times in good and poor readers. *Vision Research, 35,* 285–291.

Wilson, H. R., Levi, D. M., Maffei, L., Rovamo, J., & Devalois, R. (1990). The perception of form: Retina to striate cortex. In L. Spillmann & J. S. Werner (Eds.), *Visual perception: The neurophysiological foundations* (pp. 231–272). San Diego, CA: Academic.

Woodworth, R. S. (1938). *Experimental psychology.* New York: Holt, Rinehart & Winston.

World Health Organization. (2004). *Fact sheet: Vision 2020: The right to sight, the global initiative for the elimination of avoidable blindness.* Retrieved April 3, 2005, from http://www.v2020.org/news/documents/FactsheetBlindness-101204.pdf

Wyszecki, G., & Stiles, W. S. (1982). *Color science: Concepts and methods, quantitative data and formula* (2nd ed.). Hoboken, NJ: Wiley.

Zrenner, E. (2002). Will retinal implants restore vision? *Science, 295,* 1022–1025.

Author Index[1]

A

Abdelmour, O., 82
Abell, A. M., 9, 36
Agner, D., 112
Aguayo, M., 116
Ahn, S. J., 6, 13, 15, 39, 51, 53, 72, 155,
        167–168, 170, 183–184
Ahumada, A. J., Jr., 40
Aitsebaomo, A. P., 82
Akutsu, H., 6, 21, 28–29, 49–50, 59, 127,
        147
Alexander, K. R., 55, 62
Alfaro, L., 124
Alho, K., 90
Amedi, A., 133
Anstis, S. M., 79
Anton, S., 51
Applegate, C. A., 85–86, 90
Aquilante, K., 113, 154
Arditi, A., 27, 94, 113, 115, 117, 150,
        154–155, 160
Ashourzadeh, A., 25–26, 40
Aten, T., 125
Atkins, P., 100, 102

B

Badcock, D., 24
Bagnoud, M., 135–136

Bailey, I. L., 19–20, 39, 50, 86, 130, 132,
        151, 163–164, 172–173, 177
Baldasare, J., 40
Bane, M. C., 6, 31, 34, 48–50, 108, 111,
        115, 173, 182–184
Barnes, G., 98
Barnett, G. D., 136
Bartels, P., 126
Beckmann, P. J., 6, 11, 44, 60, 64–65, 95,
        138, 140–141, 151–153, 171
Bedell, H. E., 66, 82, 87, 127
Berger, T. D., 60, 116
Bertera, J. H., 71
Best, M., 25
Bevers, P., 3
Bigelow, C., 47, 112–113, 120
Bihan, D. L., 97
Bioni, A., 183–184
Bismuth, N., 124
Blackwood, M., 24
Blakemore, C., 59
Boder, E., 102
Bolan, T. L., 126
Boschmann, M. C., 40–41
Bouma, H., 18, 70, 74, 81–82, 146–148
Bowers, A. R., 12, 14, 85–86, 141,
        151–153, 164
Bowers, J. S., 97
Bowles, S., 160
Bowling, A., 24

---

[1]Complete citations for the 20 articles in the *Psychophysics of Reading* series by Gordon Legge and colleagues are shown in Box 1.1 on pages 5 and 6, and represented by citations such as R1 (1985) elsewhere in the book. Whenever one of these articles is cited, all of the authors of the article are listed in this index.

Subject Index